Springer Tracts in Civil Engineering

Springer Tracts in Civil Engineering (STCE) publishes the latest developments in Civil Engineering—quickly, informally and in top quality. The series scope includes monographs, professional books, graduate textbooks and edited volumes, as well as outstanding Ph.D. theses. Its goal is to cover all the main branches of civil engineering, both theoretical and applied, including:

Construction and Structural Mechanics
Building Materials
Concrete, Steel and Timber Structures
Geotechnical Engineering
Earthquake Engineering
Coastal Engineering
Hydraulics, Hydrology and Water Resources Engineering
Environmental Engineering and Sustainability
Structural Health and Monitoring
Surveying and Geographical Information Systems
Heating, Ventilation and Air Conditioning (HVAC)
Transportation and Traffic
Risk Analysis
Safety and Security

To submit a proposal or request further information, please contact: Pierpaolo Riva at Pierpaolo.Riva@springer.com, or Li Shen at Li.Shen@springer.com

More information about this series at http://www.springer.com/series/15088

Liang-Jiu Jia · Hanbin Ge

Ultra-low-Cycle Fatigue Failure of Metal Structures under Strong Earthquakes

Springer

Liang-Jiu Jia
Tongji University
Shanghai, China

Hanbin Ge
Meijo University
Nagoya, Japan

ISSN 2366-259X ISSN 2366-2603 (electronic)
Springer Tracts in Civil Engineering
ISBN 978-981-13-2660-8 ISBN 978-981-13-2661-5 (eBook)
https://doi.org/10.1007/978-981-13-2661-5

Library of Congress Control Number: 2018957049

This Springer imprint is published by the registered company Springer Nature Singapore Pte Ltd.
The registered company address is: 152 Beach Road, #21-01/04 Gateway East, Singapore 189721, Singapore

Foreword

For steel and other metal structures, prevention of fracture failure is always a main concern for engineers and researchers. From the aspect of the loading rate, previous research is mainly focused on high-cycle fatigue and impact loadings. In terms of the fracture mode, attention is mainly paid to brittle fracture, and the so-called elasto-plastic fracture with plasticity only developing at a small region close to the crack tip. In the 1994 Northridge earthquake and the 1995 Kobe earthquake, steel structures were reported to fracture under a limited number of loading cycles with large strain amplitudes, and plasticity occurred in a large region of the members and connections. Though the members and connections ruptured finally in a somewhat brittle mode, ductile fracture was found to trigger the final brittle fracture. This fracture mode is quite different from the aforementioned ones. A lot of experimental and numerical research aiming to establish theoretical models, and corresponding numerical analysis methods to clarify the failure mechanism are necessary, which is a fundamental work in advancement of structural anti-fracture performance.

The authors of this book and their team members have done a lot of work in this field for the past eight years, and this book collects the main achievements during these years. In this book, the main contents include a post-necking modification method for the true stress–true strain data till fracture and metal cyclic plasticity model at full strain range under random loading. In addition, mesoscopic ductile fracture models based on the void growth theory are proposed for both monotonic and cyclic loading, and corresponding calibration methods for the plasticity and fracture model parameters only using tension coupon tests are also presented. These make it possible to predict crack initiation of metals, members, connections, and whole structures with good accuracy to some extent. These provide very useful tools for engineers, since one can predict locations with high potential to crack and further improve the design through analyzing the main reasons. Thus, safety and economy of structural design can be improved with the assistance of these tools. In addition, the achievements of this book can also provide reference for peer researchers.

China is now in the process of urbanization, and the construction industry will still maintain a considerable scale owing to the vast hinterland of the country. In addition, steel structures have many virtues over other structural types in the aspects of prefabrication, recycling, and reusing of the material and will increasingly become priority options of a variety of constructional structures. Under such a background, the academic achievements of this book have great potentials for engineering application.

Shanghai, China
May 2018

Yiyi Chen
Tongji University

Preface

Generally, metal structures mainly fail due to net section ductile rupture, brittle fracture, global/local buckling, or high-cycle fatigue. Ductile metal, especially mild steel, has been taken as an excellent building material with many virtues such as high strength, ductility, toughness, and good weldability. The virtues of high ductility and toughness were faithfully believed by structural engineers until the 1994 Northridge earthquake in California, USA, and the 1995 Kobe earthquake, Japan. It is a coincidence that the two earthquakes happened on the same day (17 January) in winter with similar low temperatures. The two strong earthquakes leaded to significant economic loss, and over 6000 people died during the Kobe earthquake. A large number of welded moment-resisting steel framed structures including both buildings and bridges fractured at the beam-to-column connections under the strong ground motions. This failure mode is different from the aforementioned ones, which is termed as ultra-low-cycle fatigue (ULCF) or extremely low-cycle fatigue. Different from high-cycle and low-cycle fatigue problems with a fatigue life far beyond 10^2 cycles, the ULCF problems in structural engineering are correlated with strong seismic loading, commonly with a fatigue life ranging from several to hundreds of cycles.

After the two earthquakes, the failure mechanism was investigated intensively by many steel structure researchers. It has been firstly pointed out by Prof. Kuwamura that the ULCF failure process during the two earthquakes can be divided into several stages, i.e., yielding of structural steel, ductile crack initiation due to strain concentration, stable ductile crack propagation, and final brittle fracture. It is known that both low temperature and cyclic plastic straining can lead to decrease in the toughness of structural steel. With the decreasing material toughness under cyclic loading and the increasing stress concentration due to the ductile crack propagation, structural steel under ULCF loading is apt to fail due to brittle fracture eventually.

This book aims to present experimental results and theoretical advancements on ULCF failure of metal structures under strong earthquakes, where the dominant failure mechanism is ductile fracture. It is expected that this book can benefit engineers and researchers from various disciplines, such as material, civil, and mechanical engineering. Pursuing a holistic approach, the book establishes a fundamental framework for this topic, while emphasizing the importance of theoretical analysis and experimental results in the fracture evaluation of metal structures under seismic loading. The book is intended for undergraduate and graduate students who are interested in metal structures, researchers investigating steel, aluminum and other metal structures, and structural engineers working on applications correlated with cyclic large plastic loading conditions. Chapter 1 describes ULCF failures of metal structures during previous strong earthquakes, including both metallic building structures and bridges. Chapters 2 and 3 deal with metal plasticity, which is the basis of ductile fracture theories. Chapter 2 discusses how to obtain the true stress–true strain data of metal till fracture, including both pre- and post-necking stages. Chapter 3 discusses cyclic metal plasticity models focusing on the full strain range till rupture of metal. Chapters 4–6 deal with ductile fracture of structural steels. Chapters 4 and 5, respectively, gives ductile fracture models based on the void growth theory for crack initiation and propagation under monotonic loading. Chapter 6 gives the ductile fracture model for cyclic loading cases. Chapters 7 and 8, respectively, gives some applications of the ductile fracture models to steel members and connections. Chapter 9 presents a straightforward approach to calibrate the cyclic plasticity model parameters using only representative mechanical properties of aluminum. Chapter 10 discusses ULCF fracture evaluation of aluminum and applications of the ULCF fracture models to aluminum members. Finally, Chap. 11 describes the main conclusions obtained in this book and future work in the field of ULCF fracture of metal structures. The implementation details of the plasticity model, i.e., the modified Yoshida–Uemori model, are also given in the appendix.

It is our sincere desire to thank all individuals who in one way or another helped us during the experimental and numerical studies in this book. Special thanks go to Dr. Yiyi Chen, Professor at Tongji University. Our sincere thanks go to him for the continuous supports and instructions all the way. We also want to thank Dr. Hitoshi Kuwamura, Professor at the University of Tokyo, for his contributions in many achievements of this book. The experimental supports from Dr. Tsuyoshi Koyama and Prof. Jun Iyama at the University of Tokyo are also greatly appreciated. The experimental supports from students at Tongji University, the University of Tokyo, and Meijo University are also greatly appreciated.

Shanghai, China Liang-Jiu Jia

Nagoya, Japan Hanbin Ge
May 2018

Contents

Chapter 1
Introduction

Abstract This chapter first introduces the motivation and background of this book. The definition of ultra-low-cycle fatigue is given, and the main differences from high- and low-cycle fatigue problems are explained. Typical ultra-low-cycle fatigue failure of metal structures under strong earthquakes is illustrated, such as post-buckling fracture of steel members and connections, fracture of welded beam-to-column connections. The correlation between the ultra-low-cycle fatigue and ductile fracture is described, and mesoscopic ductile fracture models are briefly reviewed. Finally, the objectives and outline of this book are presented.

1.1 Background and Motivation

1.1.1 Definition of Ultra-low-Cycle Fatigue

Metal structures can fail due to many reasons, and one of the main failure modes is fracture. Fracture also can be divided into several categories, such as brittle fracture, ductile fracture, transition between ductile fracture and brittle fracture, and fatigue fracture as illustrated in Fig. 1.1. Seismic loading is one of the main concerns in structural engineering, and these fracture modes are all possible to occur during an earthquake event, where brittle fracture and ductile fracture have been reported during past strong earthquakes, such as the 1994 Northridge earthquake (e.g., Mahin 1998) and the 1995 Kobe earthquake (e.g., AIJ 1995).

Herein, a seismic loading history with large strain amplitudes exceeding the yield limit of a metal is termed as ultra-low-cycle fatigue (ULCF) loading in this book, which is different from the high- and low-cycle fatigue problems since the failure mechanisms are different. The fatigue fracture is commonly associated with elastic or small plastic deformation, while the ULCF fracture is associated with large plastic strain amplitudes. The fatigue lives of high- and low-cycle fatigue problems are beyond and below 10^4 cycles, respectively. For the ULCF problems concerned with seismic loading, the fatigue life commonly ranges from several to hundreds of cycles.

L.-J. Jia and H. Ge, *Ultra-low-Cycle Fatigue Failure of Metal Structures under Strong Earthquakes*, Springer Tracts in Civil Engineering,
https://doi.org/10.1007/978-981-13-2661-5_1

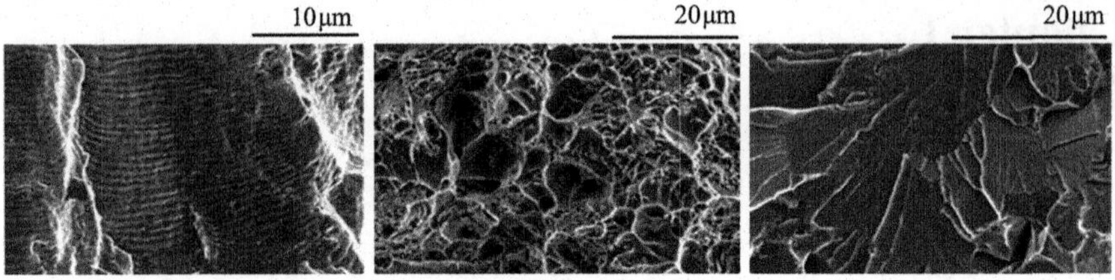

(a) Fatigue fracture surface (Striations) (b) Ductile fracture surface (Dimple pattern) (c) Brittle fracture (River pattern)

Fig. 1.1 Different fracture modes of metal structures (Liu et al. 2017)

The fracture surfaces of the aforementioned three fatigue fracture modes are also distinguished, where fatigue striations as shown in Fig. 1.1(a) can be observed in high- and low-cycle fatigue problems, and ductile dimples in ULCF ones. The fracture surface of ULCF problems is dependent on temperature, material properties, stress/strain concentration, and also the loading history. Sometimes, both ductile and brittle fracture can be observed within a single ULCF fracture surface. During a strong earthquake event, local or global buckling may first occur in some structural members such as bracings and corner columns where large deformation occurs, and ductile fracture may happen under cyclic loading with large displacement amplitudes. It is also possible that the whole ULCF fracture surface is completely ductile. The ductile fracture is mainly owing to the high toughness of the metal and also small stress/strain concentration during the ULCF loading.

From the aspect of the fracture modes, the term "ultra-low-cycle fatigue" is sometimes confusing due to its complicated failure mechanisms. The ULCF failure can be a 100% ductile fracture mode, or a brittle fracture mode induced by a preceding ductile cracking process. The ULCF fracture can also be a transition mode between the ductile and brittle fracture, or a mixture of ductile, brittle, and fatigue fracture. Therefore, it should be noted that the term ULCF is correlated with multiple meanings, and it does not indicate that the fracture surface consists of fatigue striations, which can include ductile dimples shown in Fig. 1.1(b) or a typical brittle fracture one, i.e., river pattern shown in Fig. 1.1(c).

1.1.2 Post-buckling ULCF Failure of Steel Members and Connections

During the 1995 Kobe earthquake, a number of ULCF failures of steel members were observed, which failed due to post-buckling ductile fracture. Figure 1.2 shows post-buckling ductile fracture of the bracings in a garage after cyclic buckling at the mid-length. Global buckling first occurred and led to strain concentration at the mid-length of the bracings. The cyclic loading history with large plastic strain amplitudes finally induced ductile fracture at the buckled locations. Ductile fracture of steel

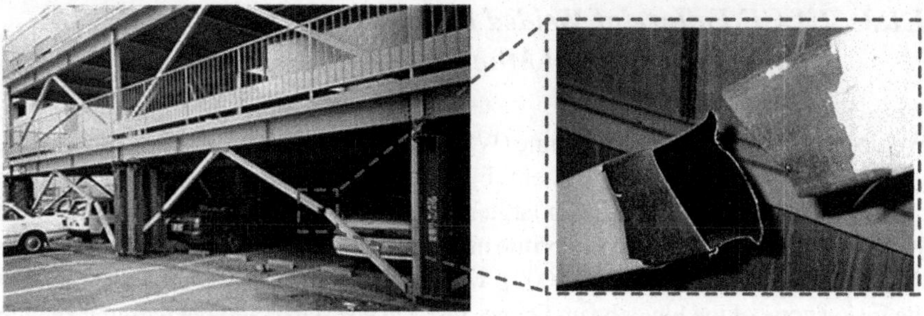

Fig. 1.2 Ductile fracture of bracings in the 1995 Kobe earthquake (Kuwamura Lab 1995)

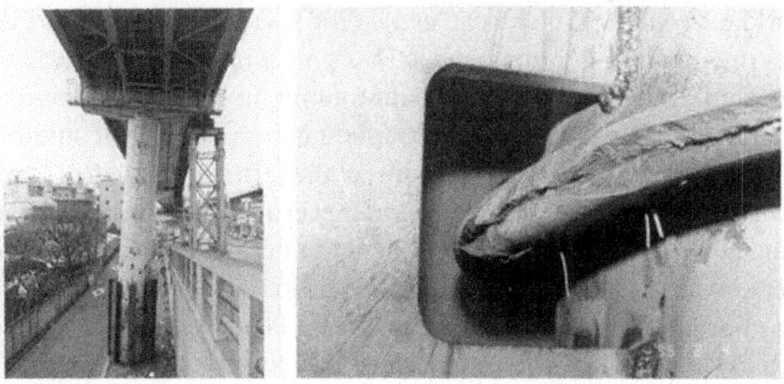

Fig. 1.3 Ductile fracture of a circular hollow section pier after elephant's foot buckling in the 1995 Kobe earthquake (Gao et al. 1998a, b)

square hollow section columns and H-beams in the 1995 Kobe earthquake was also reported (AIJ 1995). Post-buckling ductile fracture also occurred in steel bridges during the 1995 Kobe earthquake. Figure 1.3 shows post-buckling ductile fracture of a circular hollow section steel bridge pier in the elephant's foot mode under seismic loading. The reinforcing stiffeners were added to prevent collapse of the pier after the earthquake.

During the 2011 East Japan Great Earthquake, post-buckling ductile fracture failures of steel members and connections were also observed. ULCF failure of an X-shaped steel angle bracing under the seismic loading was also observed. Global buckling occurred, and strain concentrated at the buckled locations and also the bolted connections. Net section rupture finally occurred at one end of the bracing. Post-buckling ductile fracture at the connections of circular hollow section chevron bracings in a garage was also reported (Okazaki et al. 2013).

1.1.3 *ULCF Failure of Welded Beam-to-Column Connections in Steel Moment-Resisting Frames*

Welded steel moment-resisting frame (WSMRF) structures were once deemed as one of the most effective structural systems by structural engineers, while this confidence has been shaken by the fact that a large number of beam-to-column connections of WSMRF buildings fractured in a brittle mode during the 1994 Northridge earthquake (FEMA-288 1997; FEMA-350 2000), where the fracture was commonly found at the fusion zone of the lower beam flange. Just one year later, another earthquake with strong pulse-type large displacement motion hit Hyogoken-Nanbu of Japan (1995 Kobe earthquake), which also induced fracture of a large number of WSMRF buildings and bridges (AIJ 1995; Bruneau et al. 1996). Though large plastic deformation occurred at the beam-to-column connections preceding the fracture, brittle fracture finally occurred at the welds of the beam-to-column connections during the 1995 Kobe earthquake. After the two strong earthquakes, a number of studies have been conducted to clarify the failure mechanisms of the brittle fracture (e.g., Iwashita et al. 2003; Kuwamura and Yamamoto 1997; Ramirez et al. 2012; Jia and Kuwamura 2014, 2015; Xiang et al. 2017; Liu et al. 2017) and propose novel measures to improve the seismic performance, such as reduced beam section, cover plate, haunch, side-plate strengthened beam sections, tapered beam flange beam sections (e.g., Chen and Lin 2013; Chi and Uang 2002; Gilton and Uang 2002; Jones et al. 2002; Kim et al. 2002a, b; Sumner and Murray 2002). Based on experimental and numerical studies carried out after the two strong earthquakes, design recommendations have been, respectively, published to improve seismic performance of WSMRF building structures in the USA (FEMA-350 2000) and Japan (The Building Center of Japan 2003). An intensive program of research sponsored by the SAC project has been conducted to investigate various issues after the Northridge earthquake. This research included literature surveys, data collection of pre-Northridge structures, evaluation of the collected data, analytical studies of damaged and undamaged buildings, and laboratory testing of a series of full-scale beam-to-column assemblies representing typical pre-Northridge design and construction practice as well as various repair, upgrade, and alternative design details. Many of these research achievements have been included in the FEMA reports (e.g., FEMA-288 1997; FEMA-350 2000; FEMA-351 2000; FEMA-352 2000; FEMA-355a 2000). It has been found that ductile cracks first initiate and propagate stably, and brittle fracture finally occurred, leading to sudden loss of load-carrying capacity of the connections (Kuwamura and Yamamoto 1997). Due to the significant consequence of the brittle fracture, it is of great importance to investigate the stages preceding brittle fracture, i.e., ductile crack initiation and ductile crack propagation.

The ULCF fracture of the welded beam-to-column connections is different from the aforementioned steel members failing due to ductile fracture. The main factors leading to this difference include the complicated geometrical configuration of the beam-to-column connections (stress/strain concentration), relative poor ductility, and toughness of the material at the heat-affected zones (HAZs) due to the welding process.

Based on the above discussions, one has to first understand the ductile fracture mechanism to accurately evaluate the ULCF life of metal structures, since most of the ULCF failures are triggered by ductile crack initiation and propagation. Therefore, prediction of ductile fracture is of great importance to both brittle and ductile fracture problems. In addition, prediction of ductile fracture can be applied to evaluate deformation capacities and plastic energy dissipation capacities of structural steels, members, connections, and whole structures.

1.2 Ductile Fracture

1.2.1 Brief Introduction

The study on ductile fracture of uncracked metals has been studied for a relatively short history of around half a century. The mechanism of the ductile fracture commonly includes three critical stages (Anderson 2005).

(1) Void nucleation at an impurity or second-phase particle by either interface decohesion or particle cracking;
(2) Void growth depending on both the equivalent plastic strain and hydrostatic stress;
(3) Void coalescence when the voids grow to a critical size.

Several mathematical models for void growth have been proposed, among which the most widely referenced ones are the following four models, i.e., Gurson 1975; Mcclintock 1968; Rice and Tracey 1969; and modified Gurson model (Tvergaard 1981, 1982; Tvergaard and Needleman 1984).

McClintock, Rice, and Tracey found the important role of hydrostatic stress in the growth of micro-voids. McClintock proposed a criterion for ductile fracture by the growth of holes, and the model assumes cylindrical holes with elliptical cross section with axes parallel to the principle directions of the applied stress. The criterion of ductile fracture is that each hole touches a pair of its cell walls as the growth of the holes. A relative growth factor giving the increase in semi-axis of the hole relative to the corresponding hole spacing is proposed to define the crack initiation.

Rice and Tracey analyzed a more realistic spherical void in a remote simple tension strain rate field and found that the fracture ductility decreased rapidly with the increase in stress triaxiality. They also found that the growth rate of the micro-void was an exponential function with the same coefficient in the exponential term for a non-hardening material subjected to tension for the case of a spherical void. However, the Rice–Tracey model does not have a criterion for ductile fracture, and its main contribution is on the analysis results of the relationship between the growth rate of micro-voids and stress triaxiality.

The Gurson model is a porous plasticity constitutive model, which considers the effect of void through coupling of the yield surface of the material and the hydrostatic stress. The main characteristic of the model is that it is based on the continuum theory while the material is dependent on the hydrostatic stress. The condition of fracture is defined through a parameter called void volume fraction; i.e., crack forms when the parameter reaches a critical value. Based on experimental results, Tvergaard (1981, 1982) modified the yield function of the Gurson model by adding three model parameters. The Gurson model was found to overestimate the fracture strain, and Tvergaard and Needleman (1984) modified the model by introducing a failure point, after which the effect of hydrostatic stress on the yield surface accelerates. The modified model is usually called Gurson–Tvergaard–Needleman (GTN) model. The model can characterize void growth and coalescence, while the model has over ten parameters for a single material. It may be difficult to be applied in structural engineering, since commonly only test results of smooth coupons (or flat bars) under monotonic tension are available in practice. Another limitation is that the Gurson model and the GTN model adopt the classical von Mises material with isotropic hardening (IH) rule, which may not well describe cyclic plasticity of structural steels.

Another approach to simulate ductile fracture is continuum damage mechanics (CDM). The CDM is initiated from the concept of a macroscopic damage variable proposed by Kachanov (1958). Chaboche (1984) and Lemaitre (1985) then formulated the constitutive equations of CDM in the framework of thermodynamics, which provides scientific basis for the theory. Lemaitre proposed a CDM model for ductile fracture based on the concept of effective stress, which requires identification of three model parameters from the test results. He assumed that macroscopic crack initiated when the damage variable, D, reached a critical value. A method based on decrease of the Young's modulus to identify D was also proposed in his paper. To determine D, several cycles of loading and unloading tests are necessary, which is also commonly difficult to obtain in practice for structural engineers.

There are also many empirical fracture models proposed, which define scalar variables named "damage indicators," and fracture of materials is assumed to occur when the damage indicator reaches a critical value. The simplest empirical fracture model is the constant fracture strain rule, and fracture is assumed to happen when the equivalent plastic strain reaches a critical value. As it is known that the stress triaxiality plays an important role in ductile fracture of metals, many empirical fracture models include the stress triaxiality in the damage indicator (e.g., Bai et al. 2006; Bao and Wierzbicki 2004; Johnson and Cook 1985; Norris et al. 1977; Marino et al. 1985; Oyane et al. 1980). More recently, more sophisticated fracture models considering the effect of the Lode angle were also proposed (e.g., Bai and Wierzbicki 2008). However, the aforementioned fracture models are either too complicated to apply to structural engineering or the calibrations of the model parameters require a series of specimens with special configurations. Another problem is that the models are commonly for the ductile fracture under monotonic loading, and the case under ULCF loading is seldom studied, which is of great importance in structural engineering. Limited study on ductile fracture of metals using smoothly notched coupons under cyclic loading can be found (e.g., Bai et al. 2006; Pirondi and Bonora 2003).

1.2.2 *Study of Ductile Fracture in Structural Engineering*

The study of ductile fracture in structural engineering insofar is not so extensive, and the research on this field starts from around the 1990s. The initial research is mainly focused on brittle fracture of structural steels and connections, since brittle fracture of a large number of welded beam-to-column connections in steel-framed buildings has been found during the Northridge earthquake and the Kobe earthquake. It has been found that the brittle fracture of the welded beam-to-column connections is triggered by a ductile crack at a hot spot at the weld toes after sustaining a large amount of plastic strain (Kuwamura 1998). An empirical ductile fracture model was proposed by Kuwamura and Yamamoto (1997) based on a series of notched coupon tests on several structural steels under monotonic tension. However, no rules for the case under cyclic loading are proposed. Chi et al. (2006) proposed an empirical fracture model for structural steels under monotonic loading. Kanvinde and Deierlein (2006) compared results of the model with an empirical void growth model based on the Rice–Tracey model. The model was extended to the prediction of ductile fracture under ULCF loading, and the model was named cyclic void growth model (CVGM), which has two model parameters (Kanvinde and Deierlein 2007). The calibration of the two parameters is quite complicated, and tests on a series of smoothly notched round coupons under both monotonic and cyclic loadings are required, which are often impossible to carry out due to the limitation of the configuration of the specimens. For example, the thickness of thin-walled hollow section members commonly is small, and it is often impossible to manufacture smoothly notched round coupons from the members. On the other hand, it is often impossible to carry out cyclic loading of coupons due to limitation of loading devices and constructional cost in practice. Another limitation of the model is that the calculation of the damage indicator is quite complicated, which makes the implementation of the model difficult for numerical simulations. The CVGM was applied to predict the ductile fracture of steel column base plate connections by Myers et al. (2009) and steel bracings by Fell et al. (2009). In addition, it is reported that the calibration of the model parameters of the CVGM has some difficulties, and the cyclic plasticity model used in the simulation cannot well predict the cyclic behaviors of the specimens and the connections.

1.3 Objectives and Approach

This book is focused on ductile fracture under the ULCF loading histories. The available methods currently to evaluate ductile fracture of structural steels have several limitations, e.g.,

(1) Lack of proper plasticity models for ductile metals at large plastic strain ranges close to fracture of the materials.
(2) Too many plasticity and fracture model parameters to calibrate.

(3) No standard procedures to calibrate model parameters of either plasticity models or fracture models.
(4) Coupons with special configurations are necessary to calibrate the model parameters.
(5) Cyclic tests of coupons are required to calibrate the model parameters of either plasticity models or fracture models.

This book aims to solve the aforementioned limitations and finally establishes a standard and straightforward procedure to simulate ductile fracture of steel and aluminum structures using only simple tension coupon tests.

This book can be divided into five major parts. The first part built on Chaps. 2 and 3 is devoted to study the mechanical behaviors of structural steels at large plastic strain ranges till fracture and to find or construct proper plasticity models. The approach is characterized by calibrating the plasticity model parameters only from a simple coupon test. The second part of this book is assigned to Chaps. 4–6, in that the mechanisms of ductile fracture of structural steels under monotonic and ULCF loadings are studied and proper fracture models with simple formations are established, to meet the requirement that the fracture model parameters can be easily calibrated from coupon tests. The proposed fracture models are validated by monotonic tension tests as well as tension and compression cyclic tests of smooth and notched coupons. The third part, i.e., Chaps. 7 and 8, gives an applications of the aforementioned plasticity models and ductile fracture models to steel members and connections. The fourth part consisting of Chaps. 9 and 10 is devoted to cyclic plasticity and fracture evaluation of aluminum structural members under ULCF loading. Chapter 11, that is the final part, gives the main conclusions of this book and future study in the field of ULCF failure for metal structures.

1.4 Outline of This Book

This chapter introduces the background of ULCF problem in structural engineering and gives a brief literature review of the corresponding research on ULCF failure of metal structures.

Chapter 2 describes true stress–true strain behavior of structural steels at large plastic strain ranges after necking initiates. Monotonic coupon tests are carried out, and a simple method to obtain the true stress and true strain during the post-necking stage is proposed.

Chapter 3 describes cyclic coupon tests on structural steels using hourglass-type specimens, and the capacities and limitations of several classical plasticity models for metals are studied. Proper refinements are carried out for a selected plasticity model to well simulate the cyclic plasticity of structural steel till fracture. The relationship between plasticity behaviors under monotonic and cyclic loading is also studied to calibrate the model parameters of the plasticity models with only a simple coupon test.

Chapter 4 covers ductile crack initiation mechanism of structural steels under monotonic loading. A simple fracture model considering crack initiation is established, and the model is also calibrated by available monotonic test results of notched coupons of several structural steels. A simple approach to calibrate the model parameter for the fracture model with the crack initiation rule using only tensile coupon tests is also given.

Chapter 5 presents a ductile fracture model with both crack initiation and propagation rules under monotonic loading. The crack propagation rule has a single parameter in terms of energy. A calibration method for the crack propagation parameter by tension coupon tests is also proposed. Validities of the fracture model and corresponding calibration method are verified through comparison between experimental and numerical studies on Chinese high-strength steel Q460.

Chapter 6 describes the mechanism of ductile fracture of structural steels under ULCF loading, and a cyclic fracture model is proposed to evaluate ductile fracture of structural steels under a variety of ULCF loading histories. The cyclic fracture model is also coincident with the monotonic fracture model for the case of monotonic loading. The cyclic fracture model is then validated by comparison between the cyclic coupon tests carried out in Chap. 3 and the corresponding numerical simulations. The model parameters of the plasticity models and the cyclic fracture model are all calibrated using coupon tests.

Chapter 7 gives an application of the plasticity model and the cyclic fracture model to post-buckling fracture of steel square hollow section (SHS) stub columns under ULCF loadings. Experimental tests on heat-treated steel SHS stub columns with different width-to-thickness ratios are carried out under four different loading histories. The buckling and fracture modes of the specimens are discussed. Numerical fracture simulations of the specimens are conducted. The comparison between the experimental and numerical results validates the applicability of the plasticity model and the cyclic fracture model.

Chapter 8 gives an application of the plasticity model and the cyclic fracture model to post-buckling fracture of thin-walled beam-to-column connections under cyclic loading. The connections mainly fail due to post-buckling ductile fracture at the base metal of the thin-walled panel zones. Numerical simulations using the plasticity model and the cyclic fracture model can well evaluate the crack initiation, and the crack propagation can also be evaluated with acceptable accuracy. Parametric analysis is also conducted for the connections with different configurations.

Chapter 9 describes cyclic plasticity of aluminum under ULCF loading. Experiments on widely employed aluminum AL6061-T6 using double-edge-notched (DEN) coupons are conducted. A straightforward method to calibrate the classical combined hardening model, i.e., the Chaboche model, is proposed only using representative mechanical properties of the material such as yield strength, tensile strength, and necking strain. The method is validated through comparison between the experimental and numerical results of the DEN coupons. Verification of the method using aluminum buckling-restrained braces is also carried out. The comparison results both prove excellent performance of the proposed calibration method for the plasticity model.

Chapter 10 describes the mechanism of ductile fracture of aluminum under ULCF loading and the corresponding evaluation method. It is found that the cyclic fracture model proposed in the previous chapter cannot be able to accurately evaluate ductile cracking of aluminum under constant-amplitude ULCF loading. A new cyclic fracture model is thus proposed to evaluate ductile crack initiation of aluminum structures. Likewise, the cyclic fracture model is also coincident with the monotonic fracture model for the case of monotonic loading. The cyclic fracture model is then validated by comparison between the cyclic test results of the DEN aluminum coupons and the corresponding numerical results.

Finally, Chapter 11 describes the main conclusions obtained in this book and corresponding limitations of the employed approaches, upon which future work in the field of ULCF failure of metal structures is summarized.

In addition, plasticity model plays an important role in the evaluation of ULCF failure, and the implementation method of the proposed modified Yoshida–Uemori model is described in the appendix for reference of the readers, where a proper algorithm is critical to ensure accuracy, efficiency, and convergency of the high nonlinear numerical fracture analyses.

References

AIJ (1995) Fracture in steel structures during a severe earthquake. Architectural Institute of Japan, Tokyo

Anderson TL (2005) Fracture mechanics: fundamentals and applications. Taylor & Francis

Bai Y et al (2006) Fracture of prismatic aluminum tubes under reverse straining. Int J Impact Eng 32:671–701

Bai Y, Wierzbicki T (2008) A new model of metal plasticity and fracture with pressure and Lode dependence. Int J Plast 24:1071–1096

Bao Y, Wierzbicki T (2004) On fracture locus in the equivalent strain and stress triaxiality space. Int J Mech Sci 46:81–98

Bruneau M et al (1996) Performance of steel bridges during the 1995 Hyogo-ken Nanbu (Kobe, Japan) earthquake. Can J Civ Eng 23:678–713

Chaboche JL (1984) Anisotropic creep damage in the framework of continuum damage mechanics. Nucl Eng Des 79:309–319

Chen C-C, Lin C-C (2013) Seismic performance of steel beam-to-column moment connections with tapered beam flanges. Eng Struct 48:588–601

Chi B, Uang C (2002) Cyclic response and design recommendations of reduced beam section moment connections with deep columns. J Struct Eng (ASCE) 128:464–473

Chi W et al (2006) Prediction of ductile fracture in steel connections using SMCS criterion. J Struct Eng (ASCE) 132:171–181

Fell B et al (2009) Experimental investigation of inelastic cyclic buckling and fracture of steel braces. J Struct Eng (ASCE) 135:19–32

FEMA-288 (1997) Background reports on metallurgy, fracture mechanics, welding, moment connections and frame systems behavior. Federal Emergency Management Agency, Washington DC

FEMA-350 (2000) Recommended seismic design criteria for new steel moment-frame buildings. Federal Emergency Management Agency, Washington DC

FEMA-351 (2000) Recommended seismic evaluation and upgrade criteria for existing welded steel moment-frame buildings. Federal Emergency Management Agency, Washington DC

FEMA-352 (2000) Recommended postearthquake evaluation and repair criteria for welded steel moment-frame buildings. Federal Emergency Management Agency, Washington DC
FEMA-355A (2000) State of the art report on base materials and fracture. Federal Emergency Management Agency, Washington DC
Gao S et al (1998a) Ductility evaluation of steel bridge piers with pipe sections. J Eng Mech (ASCE) 3(124):260–267
Gao S et al (1998b) Ductility of steel short cylinders in compression and bending. J Eng Mech (ASCE) 2(124):176–183
Gilton C, Uang CM (2002) Cyclic response and design recommendations of weak-axis reduced beam section moment connections. J Struct Eng (ASCE) 128:452–463
Gurson AL (1975) Continuum theory of ductile rupture by void nucleation and growth. Part I. Yield criteria and flow rules for porous ductile media. Brown Univ, Providence, RI (USA). Div of Engineering
Iwashita T et al (2003) Prediction of brittle fracture initiating at ends of CJP groove welded joints with defects: study into applicability of failure assessment diagram approach. Eng Struct 25:1815–1826
Jia L-J, Kuwamura H (2014) Ductile fracture simulation of structural steels under monotonic tension. J Struct Eng (ASCE) 5(140):04013115
Jia L-J, Kuwamura H (2015) Ductile fracture model for structural steel under cyclic large strain loading. J Constr Steel Res 106:110–121
Johnson GR, Cook WH (1985) Fracture characteristics of three metals subjected to various strains, strain rates, temperatures and pressures. Eng Fract Mech 21:31–48
Jones S et al (2002) Experimental evaluation of cyclically loaded reduced beam section moment connections. J Struct Eng (ASCE) 128:441–451
Kachanov LM (1958) Time of the rupture process under creep conditions. Izvestiya Akademii Nauk SSSR Otdelenie Tekniches 8:26–31
Kanvinde A, Deierlein G (2006) The void growth model and the stress modified critical strain model to predict ductile fracture in structural steels. J Struct Eng (ASCE) 132:1907–1918
Kanvinde A, Deierlein G (2007) Cyclic void growth model to assess ductile fracture initiation in structural steels due to ultra low cycle fatigue. J Eng Mech (ASCE) 133:701–712
Kim T et al (2002a) Cover-plate and flange-plate steel moment-resisting connections. J Struct Eng (ASCE) 128:474–482
Kim T et al (2002b) Experimental evaluation of plate-reinforced steel moment-resisting connections. J Struct Eng (ASCE) 128:483–491
Kuwamura H (1998) Fracture of steel during an earthquake—state-of-the-art in Japan. Eng Struct 20:310–322
Kuwamura H, Yamamoto K (1997) Ductile crack as trigger of brittle fracture in steel. J Struct Eng (ASCE) 123:729–735
Lab Kuwamura (1995) Field survey report on structural damage during the 1995 Hyogoken-Nanbu Earthquake. Kuwamura Lab, School of Engineering, The Univ of Tokyo, Tokyo
Lemaitre J (1985) A continuum damage mechanics model for ductile fracture. J Eng Mater Technol 107:83–89
Liu Y et al (2017) Ductile-fatigue transition fracture mode of welded T-joints under quasi-static cyclic large plastic strain loading. Eng Fract Mech 176:38–60
Mahin SA (1998) Lessons from damage to steel buildings during the Northridge earthquake. Eng Struct 20:261–270
Marino B et al (1985) Experimental study of cavity growth in ductile rupture. Eng Fract Mech 22:989–996
McClintock FA (1968) A criterion for ductile fracture by the growth of holes. J Appl Mech 35:363–371
Myers AT et al (2009) Effect of weld details on the ductility of steel column baseplate connections. J Constr Steel Res 65:1366–1373

Norris DM et al (1977) A Plastic-strain, mean-stress criterion for ductile fracture. J Eng Mater Technol 100:279–286
Okazaki T et al (2013) Damage to steel buildings observed after the 2011 Tohoku-Oki Earthquake. Earthq Spectra 29:S219–S243
Oyane M et al (1980) Criteria for ductile fracture and their applications. J Mech Work Tech 4:65–81
Pirondi A, Bonora N (2003) Modeling ductile damage under fully reversed cycling. Comput Mater Sci 26:129–141
Ramirez CM et al (2012) Fragility functions for pre-Northridge welded steel moment-resisting beam-to-column connections. Eng Struct 45:574–584
Rice JR, Tracey DM (1969) On the ductile enlargement of voids in triaxial stress fields. J Mech Phys Solids 17:201–217
Sumner E, Murray T (2002) Behavior of extended end-plate moment connections subject to cyclic loading. J Struct Eng (ASCE) 128:501–508
The Building Center of Japan (2003) Guidelines for prevention of brittle fracture at the beam ends of welded beam-to-column connections in steel frames. The Building Center of Japan, Tokyo
Tvergaard V (1981) Influence of voids on shear band instabilities under plane strain conditions. Int J Fract 17:389–407
Tvergaard V (1982) On localization in ductile materials containing spherical voids. Int J Fract 18:237–252
Tvergaard V, Needleman A (1984) Analysis of the cup-cone fracture in a round tensile bar. Acta Metall 32:157–169
Xiang P et al (2017) Ductile cracking simulation of uncracked high strength steel using an energy approach. J Constr Steel Res 138:117–130

Chapter 2
Stress–Strain Behaviors in Large Plastic Strain Ranges under Monotonic Loading

Abstract For ultra-low-cycle fatigue problems, it is a prerequisite to first understand the plasticity behaviors of metal. For monotonic loading, the true stress–true strain data till fracture is of great importance for ductile fracture evaluation. Post-necking modification for the true stress–true strain data is required since the stress state after necking is not uniaxial. In this chapter, a modified weighted average method is proposed and compared with other two methods. The applicability of the method for structural steel is also proved through tension coupon tests and corresponding numerical simulation.

2.1 Introduction

Finite element method (FEM) has become a powerful tool in solving many elasto-plastic engineering problems. For elasto-plastic numerical simulation, the FEM requires the uniaxial true stress–true strain data, which is commonly obtained from a tension coupon test. However, there is a difficulty in obtaining the uniaxial true stress–true strain data after necking initiation for tension coupons, since the stress state after necking initiation is triaxial. Though the true stress–true strain relationship after necking is very important, especially in the analysis of ductile fracture, a limited number of empirical methods have been proposed, e.g., Bridgman's modification method (Bridgman 1952) and the weighted average (WA) method (Ling 1996). However, they are either not applicable in practice because of complexity or poor accuracy. In this chapter, a modification of the WA method, herein named as modified weighted average (MWA) method, is proposed based on three assumptions (Jia and Kuwamura 2014). In this chapter, the MWA method will be found capable of predicting the global behavior of monotonic coupon tests after necking and will be utilized in subsequent chapters to establish the basic true stress–true strain curves for cyclic plasticity analysis and fracture analysis.

L.-J. Jia and H. Ge, *Ultra-low-Cycle Fatigue Failure of Metal Structures under Strong Earthquakes*, Springer Tracts in Civil Engineering,
https://doi.org/10.1007/978-981-13-2661-5_2

2.2 True Stress–True Strain After Necking Initiation

2.2.1 Definitions of True Stress–True Strain

For a bar with a uniform cross section, the original cross-sectional area and original length are denoted by A_0 and l_0, respectively, and those of the current are A and l, respectively. Given an incremental change, dl, incremental strain, dε, is defined as

$$\mathrm{d}\varepsilon = \frac{\mathrm{d}l}{l} \tag{2.1}$$

The total strain can be obtained by integral of Eq. (2.1).

$$\varepsilon = \int_0^{\varepsilon} \mathrm{d}\varepsilon = \int_{l_0}^{l} \frac{\mathrm{d}l}{l} = \ln \frac{l}{l_0} \tag{2.2}$$

The strain defined by Eq. (2.2) is called true strain, natural strain, or logarithmic strain.

The engineering strain e is defined as

$$e = \frac{l - l_0}{l_0} \tag{2.3}$$

The corresponding engineering stress is defined as

$$s = \frac{P}{A_0} \tag{2.4}$$

The true stress is defined as

$$\sigma = \frac{P}{A} \tag{2.5}$$

The engineering stress and strain are defined with reference to the undeformed configuration, while the definitions of the true stress and true strain consider the reduction of the cross-sectional area.

The above definitions are all under the assumptions of uniform stress and strain distribution under uniaxial stress state. After necking initiation, since the stress state becomes triaxial, it is impossible to measure the uniaxial true stress and true strain directly from experiments. However, the uniaxial true stress–true strain data after necking initiation is required by numerical simulations.

It is assumed that for any stress state, there is an equivalent uniaxial stress state. Then, one just needs to correlate the stresses and strains in a triaxial stress state to the uniaxial stress state. An equivalent uniaxial stress, i.e., the von Mises stress, is commonly employed for metals

$$\sigma_e = \sqrt{\frac{1}{2}[(\sigma_x - \sigma_y)^2 + (\sigma_y - \sigma_z)^2 + (\sigma_z - \sigma_x)^2 + 6(\tau_{xy}^2 + \tau_{yz}^2 + \tau_{zx}^2)]} \tag{2.6}$$

where σ_x, σ_y, and σ_z are the normal stress components, and τ_{xy}, τ_{yz}, and τ_{zx} are the shear stress components. An equivalent uniaxial strain can also be defined as

$$\varepsilon_e = \int_0^{\varepsilon_e} \mathrm{d}\varepsilon_e \tag{2.7}$$

where

$$\mathrm{d}\varepsilon_e = \sqrt{\frac{2}{9}[(\mathrm{d}\varepsilon_x - \mathrm{d}\varepsilon_y)^2 + (\mathrm{d}\varepsilon_y - \mathrm{d}\varepsilon_z)^2 + (\mathrm{d}\varepsilon_z - \mathrm{d}\varepsilon_x)^2 + 6(\mathrm{d}\gamma_{xy}^2 + \mathrm{d}\gamma_{yz}^2 + \mathrm{d}\gamma_{zx}^2)]} \tag{2.8}$$

where ε_x, ε_y, and ε_z are the normal strain components, and γ_{xy}, γ_{yz}, and γ_{zx} are the shear strain components. In large strain analysis, there are a number of different strain definitions, e.g., Almansi strain, logarithmic strain, Green strain. The different definitions give different values of strain. In metal plasticity, the logarithmic strain is commonly employed. In finite element software such as ABAQUS, strain is calculated from the deformation gradient, which is detailed in the literature (Dunne and Petrinic 2005).

After necking initiation, the equivalent stress and equivalent strain are employed for the input of true stress–true strain data, i.e.,

$$\sigma_e = \sigma, \varepsilon_e = \varepsilon. \tag{2.9}$$

2.2.2 *Criterion for Necking Initiation*

Before the analysis of the post-necking stress state, it is necessary to find out the condition of necking initiation.

Based on the assumption of volume constancy of metals, one can obtain

$$\mathrm{d}\varepsilon = -\frac{\mathrm{d}A}{A} \tag{2.10}$$

Necking is assumed to initiate at the peak of a load–deformation curve of a coupon test, where the incremental force, dP, is zero. The differential of P gives

$$\mathrm{d}P = A \cdot \mathrm{d}\sigma + \sigma \cdot \mathrm{d}A = 0 \tag{2.11}$$

Substituting Eq. (2.10) into Eq. (2.11), one can obtain the criterion for the necking initiation as

$$\sigma = \frac{\mathrm{d}\sigma}{\mathrm{d}\varepsilon} \tag{2.12}$$

which indicates that necking initiates when the current true stress is equal to the current stress gradient. The above argument is well known and commonly accepted in solid mechanics.

2.2.3 Simple Determination Method

The following will give the formulae of the commonly used method named simple determination (SD) method herein to obtain the true stress and true strain from a tension coupon test.

Under the assumption of volume preservation, i.e., $A_0 l_0 = Al$, the true stress is given as

$$\sigma = \frac{P}{A} = s \cdot (1 + e) \tag{2.13}$$

and the true strain is given as

$$\varepsilon = \mathbf{ln}\frac{l}{l_0} = 2\,\mathbf{ln}\frac{d_0}{d} = \mathbf{ln}(1 + e) \tag{2.14}$$

However, this equation holds only till necking initiation, since the deformation of a coupon will concentrate on the necked region and the region will be in triaxial stress state during the post-necking stage.

2.2.4 Weighted Average Method

The SD method for the true stress and true strain is based on the following two assumptions:

1. The strain along the coupon length is uniform.
2. The stress across the cross section is uniform.

However, the two assumptions are not valid after necking initiation, since the stress and strain across the cross section are non-uniform. The stress state within the neck will be triaxial, and the true stress obtained by Eq. (2.13) has to be modified after necking initiation. The true strain obtained by Eq. (2.14) is also not accurate.

Limited efforts have been made by previous researchers to predict the true stress within the necked region, and several methods have been proposed. The most famous one is Bridgman's modification. However, it is difficult to use since it is complicated to determine the variables required by the method and also the method is not applicable for coupons with rectangular sections.

The WA method proposed by Ling (1996) assumes an empirical lower bound and a linear upper bound for the post-necking true stress–true strain. The empirical lower bound is assumed to be a power law curve which can be extrapolated by fitting of the true stress–true strain curve before necking. However, it is found that the method sometimes fails in obtaining an accurate simulation result because the proposed lower bound overestimates the post-necking true stress for some metals and is actually not always a lower bound of a true stress–true strain curve.

2.2.5 Modified Weighted Average Method

Necking initiates when the criterion of Eq. (2.12) is satisfied. After necking initiation, the curve obtained using the SD method is not accurate. However, three assumptions can be made from the test results of mild steels (Jia and Kuwamura 2014).

1. The hardening modulus after necking is smaller than that at the point where necking initiates, which is in accordance with the assumption of the upper bound proposed by Ling. Therefore, the true stress when necking initiates is an upper bound of the hardening modulus after necking.
2. The hardening modulus after necking is larger than zero, which means zero is a lower bound of the hardening modulus after necking.
3. The true stress–true strain curve after necking is almost linear, which indicates that the hardening modulus after necking is approximately a constant. This assumption is proved to be valid for many steels at large plastic strain ranges by classical tests done by Bridgman.

According to Assumption 1, an upper bound of the true stress after necking can be given as

$$\sigma = \sigma_{\text{neck}} + \sigma_{\text{neck}}(\varepsilon - \varepsilon_{\text{neck}}) \tag{2.15}$$

where σ_{neck}, $\varepsilon_{\text{neck}}$ are the true stress and true strain when necking initiates, respectively, and σ, ε are the current true stress and true strain, respectively.

According to Assumption 2, the constant σ_{neck} is a lower bound of the true stress after necking (the hardening modulus is zero). Then according to Assumption 3, an optimal weight average factor, w, between the upper bound and the lower bound which gives the best fitting of the actual post-necking hardening modulus can be obtained. Using the weighted average factor, w, the extrapolation of the post-necking true stress can be given as

$$\sigma = \sigma_{\text{neck}} + w \cdot \sigma_{\text{neck}}(\varepsilon - \varepsilon_{\text{neck}}) \tag{2.16}$$

where σ_{neck} and $\varepsilon_{\text{neck}}$ can be easily determined by the SD method.

To obtain the optimal weighted average factor, w, one has to measure the diameter of the minimum cross section of a coupon, and several iterations are necessary to obtain a best fitting of the experimental load–deformation curve using numerical simulation. The detailed procedure of the modified weighted average (MWA) method is:

Step 1: Calculate the true stress and true strain according to Eqs. (2.13) and (2.14), and obtain σ_{neck} and ε_{neck}.
Step 2: Give an initial value of w, e.g., $w=0.5$, and modify the true stress larger than σ_{neck} according to Eq. (2.16).
Step 3: Carry out a numerical simulation of the coupon test using the material data obtained in Step 2, and compare the experimental load–deformation curve with the simulation result.
Step 4: If the numerical result in Step 3 compares well with the test result, then the w is the optimal value. If the comparison result is not good, go back to Step 2 and give a new w based on the comparison result (if the load–deformation curve of the numerical result is larger than that of the experiment, then give a smaller w, and vice versa), and repeat the loop until the comparison result is acceptable.

2.3 Experiment

2.3.1 Material

The material of the coupons is SS400 (JIS G3101 2015), and the material properties and chemical composition provided by the maker are listed in Tables 2.1 and 2.2, respectively.

Table 2.1 Material properties of the steel (mill sheet)

Grade	Yield stress (N/mm^2)	Ultimate stress (N/mm^2)	Elongation (%)
SS400	265	462	35

Table 2.2 Chemical composition of the steel (mass%)

Grade	C	Si	Mn	P	S
SS400	0.2	0.15	0.8	0.18	0.04

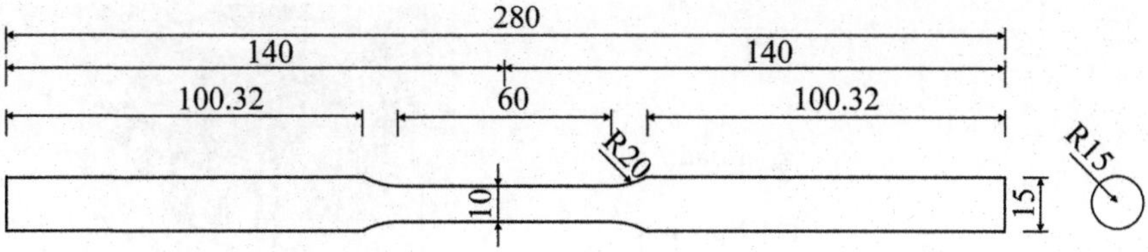

Fig. 2.1 Configuration of tensile coupon tests

2.3.2 Configuration of Coupons

To validate the proposed MWA method, two tension coupons are designed according to JIS-Z2201 (2005) as shown in Fig. 2.1. The minimum section is selected to be 10 mm considering the capacity of the test machine. For the coupon test, the parallel part is long enough to ensure that the stress and strain are uniform across the minimum section before necking initiation. Two coupons were taken from a steel plate of 65 mm thick. As is known, the impurity substance usually accumulates near the neutral surface of thick steel plates, so the coupons were taken from the part near the plate surface to avoid the influence.

2.3.3 Test Setup

The coupons were tested in the Steel Structure Laboratory of the Department of Architecture, the University of Tokyo. The test setup is shown in Fig. 2.2, where the coupon was loaded at the ends by two loading heads. The minimum diameters of the coupons were measured using a digital micrometer Keyence TM-3000. The elongations of the coupons were not measured because of difficulty in the test setup. Strain gauges were also used to obtain more accurate strain results. However, they failed prematurely (up to around 20%) before rupture of the coupons.

2.3.4 Experimental Results

The load–deformation curves of coupon *KAcoupon1* and *KAcoupon2* are shown in Fig. 2.3. It can be found that they are approximately the same except that the noise of *KAcoupon1* is much larger than that of *KAcoupon2*. Therefore, the test data of *KAcoupon2* is adopted in the present study. The MWA method is illustrated by the test result of *KAcoupon2* in Fig. 2.4. It can be found that the true stress–true strain curve obtained by the SD method lies between the lower and upper bounds of the MWA method, and the post-necking stress–strain curve is approximately linear.

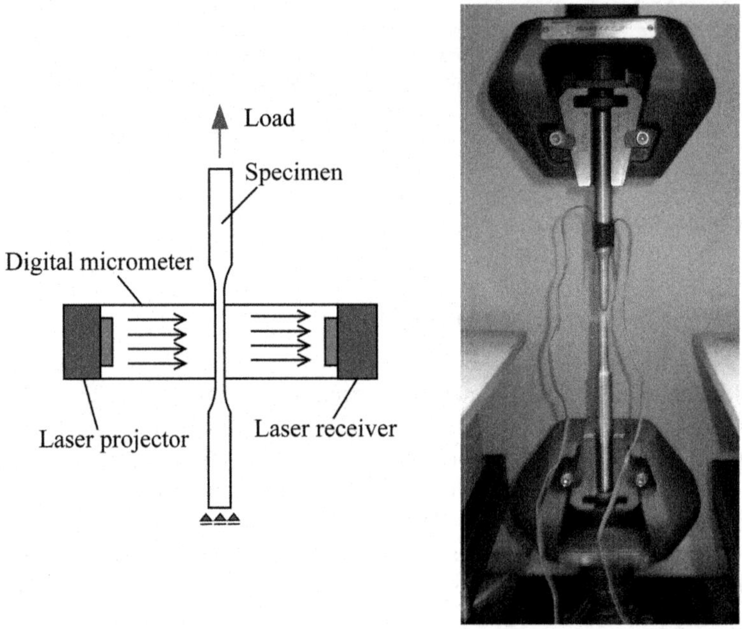

Fig. 2.2 Test setup of coupon tests

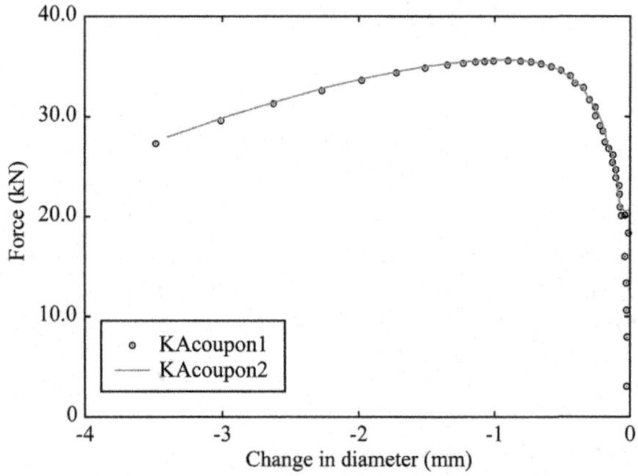

Fig. 2.3 Load change in diameter curves of tension coupon tests

2.4 Simulation

The coupons are simulated with an axisymmetric half model using CAX8 elements in ABAQUS (2010) as shown in Fig. 2.5. Only the central portion with a uniform cross section is simulated, and symmetric boundary conditions are applied to the left end with the right end subjected to displacement loading. The nonlinear isotropic hardening model in ABAQUS is employed for the simulation. The yield criterion is

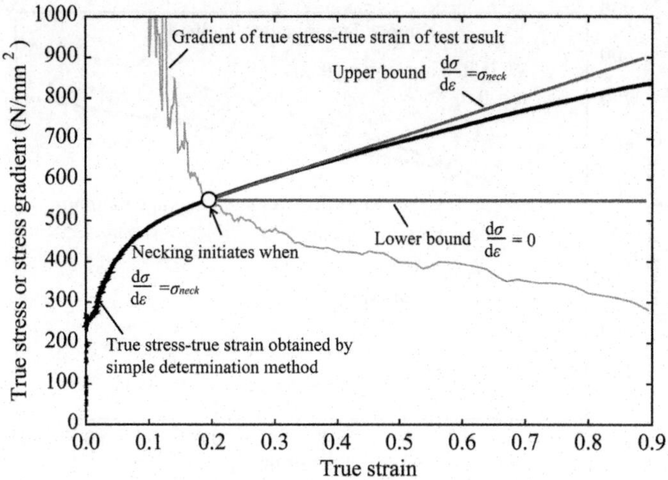

Fig. 2.4 Stress gradient of *KAcoupon2*

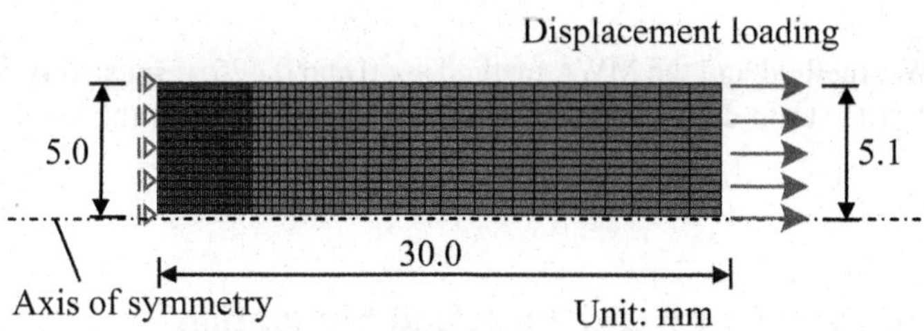

Fig. 2.5 Mesh of finite element model

Mises yield function, and the flow rule is the associated flow rule, which is commonly accepted for metals.

For an ideal symmetric numerical model with the same nominal geometry as the coupons, it is difficult to predict the post-necking behavior since the numerical model will elongate without necking. To simulate necking, a small geometric slope is given to the model along the longitudinal direction of the coupon as illustrated in Fig. 2.5. The radius of the central section of the coupon is 5.0 mm, while the radius of the end of the flat part is increased to 5.1 mm. Since the slope is only 0.33%, which is very small, it will not have an apparent effect on the global behavior. It should be noted that the stress state of the necked part after necking initiation is triaxial, and the numerical simulation is not the copy of the experimental stress–strain curve.

The stress–strain curves obtained by the SD method, the WA method, and the MWA method are compared in Fig. 2.6. The true stress, σ_{neck}, and the true strain, ε_{neck}, when necking occurs are 558 MPa and 0.20, respectively. The weighted factors,

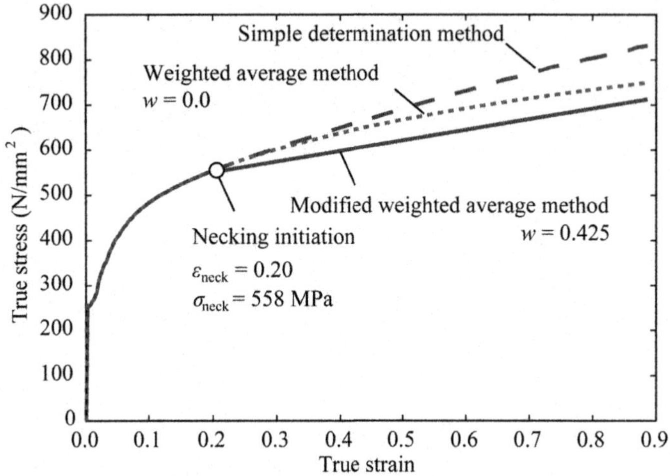

Fig. 2.6 Comparison of true stress and true strain using three methods

w, of the WA method and the MWA method are 0 and 0.425, respectively, where the true stress–true strain data evaluated by the WA method is just the assumed lower bound.

2.5 Comparison of Experiment and Simulation

The load–deformation curves of the test and numerical results using the true stress–true strain data obtained by the three methods are compared as shown in Fig. 2.7. It can be found that the results of the three methods compare well with the test results before necking initiation, while the SD method and the WA method over-estimate the post-necking forces. The MWA method proposed in the present study gives an accurate result even at extremely large plastic strain ranges. The differences among the three methods become larger as strain increases.

According to the classical hydrostatic monotonic tension coupon tests on several steels done by Bridgman (1952), it has been found that the strains where necking initiates for the tested mild steels are almost the same as 20%, and the post-necking stress–strain relationship is approximately linear. It is also commonly accepted that mild steels will go on hardening after necking initiation. Accordingly, the three assumptions of the MWA method are commonly accepted, and it is reasonable to believe that the method is applicable for a variety of mild steels.

According to the experimental study in the literature (Besson et al. 2001), the effect of void on stress–strain behavior becomes apparent only in strain ranges close to the failure point of a monotonic tension coupon. Therefore, it is rational to use the MWA method to predict the stress–strain behavior of structural steel till fracture.

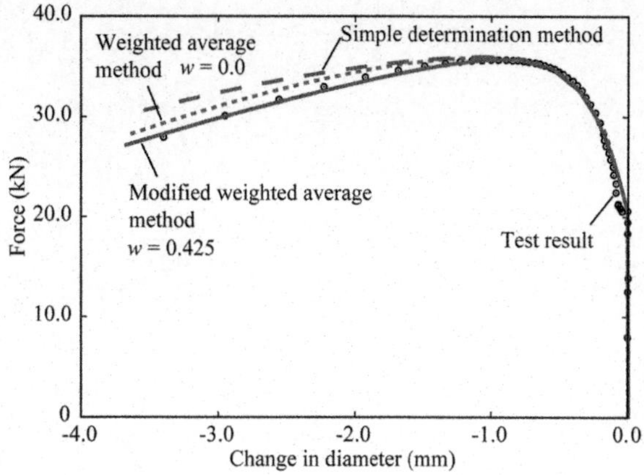

Fig. 2.7 Comparison between test and numerical results

2.6 Summaries

Monotonic tension tests on coupons of mild steels are carried out, and the true stress–true strain behaviors at large plastic strain are studied. A modified weighted average method is proposed to modify the post-necking true stress–true strain. The method is straightforward and can well evaluate the stress–strain behaviors at large plastic strain.

References

ABAQUS (2010) ABAQUS standard manual (Version 6.10). Karlsson & Sorensen Inc, Hibbitt Pawtucket (RI, USA)

Besson J et al (2001) Modeling of crack growth in round bars and plane strain specimens. Int J Solids Struct 38:8259–8284

Bridgman PW (1952) Studies in large plastic flow and fracture. McGraw-Hill New York

Dunne F, Petrinic N (2005) Introduction to computational plasticity. Oxford University Press

G3101 J (2015) Rolled steels for general structure. Japanese Industrial Standards Committee, Tokyo

Jia L-J, Kuwamura H (2014) Ductile fracture simulation of structural steels under monotonic tension. J Struct Eng (ASCE) 140 5:04013115

JIS-Z2201 (2005) Test pieces for tensile test for metallic materials. JIS, Tokyo

Ling Y (1996) Uniaxial true stress-strain after necking. AMP J Technol 5:37–48

Chapter 3
Stress–Strain Behaviors in Large Plastic Strain Ranges under Cyclic Loading

Abstract Before studying on ultra-low-cycle fatigue problems, one has to first understand cyclic plasticity of metal. For ductile fracture under ultra-low-cycle fatigue loading, the corresponding plastic strain level is much larger than conventional engineering problems, where extremely large plastic strain is of concern. In this chapter, a number of classical rate-independent cyclic plasticity models for metal are investigated. Formations and limitations of the plasticity models are introduced. A modified Yoshida–Uemori model is proposed to solve the aforementioned limitations. In addition, a method to calibrate the plasticity model parameters is also proposed, where only tension coupon test results are required. Experimental tests on several specimens are conducted, and applicability of the models is compared. Applicability of the aforementioned models to pre-strained structural steel is also investigated.

3.1 Introduction

Modeling cyclic plasticity of mild steels is of great importance in the analyses of steel structures subjected to cyclic loading. For most practical applications in structural engineering, only test results of monotonic coupon tests can be obtained, which are then used to calibrate the model parameters of cyclic plasticity models. However, simulation results are usually unsatisfactory due to the limitations of available data and capacity of plasticity models.

In this chapter, a blind test procedure is devised with seven test specimens and two coupon tests to find a reasonable methodology of calibrating three plasticity models using only tension coupon test result, i.e., the Prager model, the Chaboche models, and the Yoshida–Uemori model. The blind test is a scientific technique, for which the people involved in the test are kept unknown with certain information which may lead to bias on the test results. For natural science, it is of great importance to carry out a blind test because the methodology can prevent the test results from being affected by subjective consciousness. Therefore, the calibration of the model parameters for the simulation of each specimen is based on the blind test concept, which prevents the

L.-J. Jia and H. Ge, *Ultra-low-Cycle Fatigue Failure of Metal Structures under Strong Earthquakes*, Springer Tracts in Civil Engineering,
https://doi.org/10.1007/978-981-13-2661-5_3

analyst from knowing the corresponding test results of the simulation and ensures a subjective procedure. Once the simulation results can agree well with the test results, then a fixed standard methodology for the calibration of model parameters will be found, which can also be applied to other test results. Based on the blind test, it is found that the Chaboche model gives good simulation results for most of the specimens, and the Yoshida–Uemori model can also give good prediction results after proper modifications (Jia and Kuwamura 2014a).

3.2 Mathematical Models

3.2.1 Review of Mathematical Theory of Plasticity

(a) *A Short Introduction of Rate-Independent Metal Plasticity*
Generally, to describe cyclic behaviors of metal plasticity, it is necessary to give

1. a yield function which defines the yield condition;
2. a flow rule which defines the evolution law of plastic strain;
3. a hardening rule which specifies the post-yielding behaviors.

For metal plasticity, the Mises yield function and the associated flow rule are generally accepted and are herein employed. In this chapter, the main concern is the hardening rule. There are two basic hardening rules for metal plasticity, i.e., isotropic hardening (IH) and kinematic hardening (KH). Since the IH rule alone cannot describe the Bauschinger effect that appears in the cyclic behaviors of metals, most of the hardening rules are developed based on a KH rule or combination of the two basic rules. Since the yield function and the flow rule are determined, the terminology "hardening model" in the following denotes a hardening rule with the Mises yield function and the associated flow rule.

Isotropic hardening is introduced through a scalar R. As R increases, the elastic domain also increases. R is given as a function of equivalent plastic strain as

$$R = R(\varepsilon_{\mathrm{eq}}^{p}) \tag{3.1}$$

where $\varepsilon_{\mathrm{eq}}^{p}$ is equivalent plastic strain. The most commonly used IH model is the linear IH model as shown in Fig. 3.1, where R is a linear function of $\varepsilon_{\mathrm{eq}}^{p}$, and the function $R(\varepsilon_{\mathrm{eq}}^{p})$ also can be a nonlinear function.

For the available hardening models for metals, most of them are developed from the simplest KH model termed as Prager model (Prager 1949), where the evolution of the KH is defined as a variable named backstress, $\boldsymbol{\alpha}$, which is a linear function of the plastic strain as follows

$$\mathrm{d}\boldsymbol{\alpha} = C_0 \mathrm{d}\boldsymbol{\varepsilon}_{\mathbf{p}} \tag{3.2}$$

Fig. 3.1 Illustration of linear isotropic hardening model

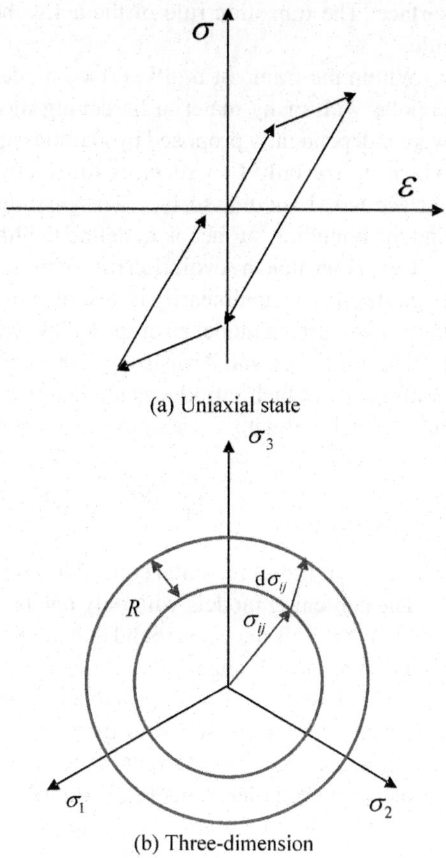

where C_0 is a model parameter, and $d\boldsymbol{\varepsilon}_\mathbf{p}$ is the incremental plastic strain.

The Prager model with Mises yield function can be expressed as

$$f = \sqrt{\frac{3}{2}(\mathbf{S} - \boldsymbol{\alpha}) : (\mathbf{S} - \boldsymbol{\alpha})} - \sigma_{y0} = 0 \tag{3.3}$$

where f is the yield function, $\mathbf{S}$ and $\boldsymbol{\alpha}$ are deviatoric stress and backstress, respectively, and σ_{y0} is the initial yield stress.

However, the linear KH is usually too rough in predicting the KH component of material hardening. A series of commonly used material hardening models are developed from the multi-surface model initially proposed by (Iwan 1967; Mróz 1967), which utilize a family of surfaces in the stress space. However, one of the main shortcomings of the models is the evolution rules of the multi-surface models, which are too complicated to apply. The yield surface is the inner smallest surface and can move within a series of bounding surfaces. When the yield surface meets the subsequent bounding surface, then the bounding surface is defined as the active surface. The transition rule of the active bounding surface is given by the Mróz's rule.

Within the frame of multi-surface models, the two-surface model is of the most popular with many material hardening models proposed. The two-surface models were independently proposed by Dafalias and Popov (1975, 1976) and Krieg (1975), which utilize only two surfaces to describe the nonlinear behaviors, i.e., a yield surface and a bounding surface. The yield surface is used to define the yield condition, and the bounding surface is to define the limiting state of the stress state.

It is known that the evolution rule of most metals is not linear except at significantly large strain. The nonlinearity is described by a continuously varying plastic modulus in the two-surface model as proposed by Dafalias and Popov (1975, 1976) and Krieg (1975). Frederick and Armstrong (2007) proposed another approach to describe the nonlinearity, which introduces a relaxation term into the backstress. The linear KH rule is modified with a relaxation term called *dynamic recovery* as follows

$$\mathrm{d}\boldsymbol{\alpha} = \frac{2}{3}C\mathrm{d}\boldsymbol{\varepsilon}_{\mathbf{p}} - \gamma \cdot \boldsymbol{\alpha} \cdot \mathrm{d}\varepsilon_{\mathrm{eq}}^{p} \tag{3.4}$$

where $\mathrm{d}\varepsilon_{\mathrm{eq}}^{p}$ is the incremental equivalent plastic strain, and C, γ are model parameters.

The hardening models with only one backstress cannot simulate the stress–strain relationship well since it is found that the KH component has different evolution rates at different strain ranges. The Armstrong–Frederick model was then generalized (Chaboche and Dang 1979; Chaboche 1986) by superimposing a combination of independent backstresses with different formations, which can well describe the nonlinear plasticity of different materials in different strain ranges. The generalized Armstrong–Frederick model is commonly termed as the Chaboche model. A KH rule with several backstresses is proposed by Chaboche and Dang (1979) as shown in Eq. (3.5) which gives a much better description of the transition of elastic domain and onset of plasticity.

$$\boldsymbol{\alpha} = \sum_{i}^{n} \boldsymbol{\alpha}_{\mathbf{i}}; \quad \mathrm{d}\boldsymbol{\alpha}_{\mathbf{i}} = \frac{2}{3}C_i\mathrm{d}\boldsymbol{\varepsilon}_{\mathbf{p}} - \gamma_i \cdot \boldsymbol{\alpha}_{\mathbf{i}} \cdot \mathrm{d}\varepsilon_{\mathrm{eq}}^{p} \tag{3.5}$$

where $\boldsymbol{\alpha}_{\mathbf{i}}$ and n are the ith backstress and total number of backstresses, respectively. The original Chaboche model is a nonlinear KH model, and the model is also modified by addition of the IH components. In this chapter, the original Chaboche model is named as Chaboche model without IH, and the one with IH component is named as Chaboche model with IH.

It has been found in the strain controlled test in many metal materials, e.g., 316 stainless steel (Chaboche and Dang 1979) that subsequent cyclic hardening will happen if a larger strain level is applied to a material the stress of which has already stabilized at a smaller strain range. This effect is called *plastic range memorization effect*, and (Chaboche and Dang 1979) firstly proposed a solution which adopts a state variable depending on the plastic strain. The solution is then generalized (Ohno 1982; Ohno and Kachi 1986) which introduces a non-hardening surface in the plastic strain surface. A modified two-surface model with a non-hardening surface was proposed for structural steels, which is applicable to predict cyclic plasticity within the yield plateau (Shen et al. 1995; Mamaghani et al. 1995). A different solution to consider the effect is proposed by Yoshida and Uemori (2002) which defines a non-IH surface in the stress space. The Yoshida–Uemori model also adopts a different coupled evolution rule for the KH and IH components from the Ohno model.

(b) *Adopted Models in This Chapter*
In this chapter, three cyclic plasticity models including the Prager model, the most commonly used Chaboche models (Chaboche model without IH and Chaboche model with IH), and the newly proposed Yoshida–Uemori model are selected to simulate the cyclic plasticity of structural steels. Proper modifications of the Yoshida–Uemori model are made, and the modified model is also calibrated by comparison between experimental and numerical results. The model parameters of all the models are calibrated from only tension coupon test results. The constitutive equations of each model are detailed in the following sessions.

3.2.2 Prager Model

This model is one of the most basic models to simulate the Bauschinger effect of metals, which is included in ABAQUS termed as *linear kinematic model*. Equation (3.2) can be written as follows if the Mises yield function is used

$$\mathrm{d}\boldsymbol{\alpha} = C_0(\boldsymbol{\sigma} - \boldsymbol{\alpha})\mathrm{d}\varepsilon_{\mathrm{eq}}^{p} \tag{3.6}$$

where C_0 is a model parameter, and the initial value of $\boldsymbol{\alpha}$ is zero.

The hardening model is illustrated in Fig. 3.2. It is pointed out that the Prager model is only reasonable within a strain range of 5% (ABAQUS 2010).

3.2.3 Chaboche Models Without and with IH

The Chaboche model is one of the most popular models which has been incorporated in most of finite element (FE) software, and it is also included in ABAQUS. However, it is also pointed out that the model parameters are too many and a series of cyclic tests are necessary to calibrate the model parameters.

In this chapter, the aforementioned two Chaboche models are employed, i.e., the Chaboche model without IH and the Chaboche model with IH. The Chaboche model without and with IH is named as nonlinear KH model and combined model in ABAQUS, respectively.

Fig. 3.2 Illustration of Prager model

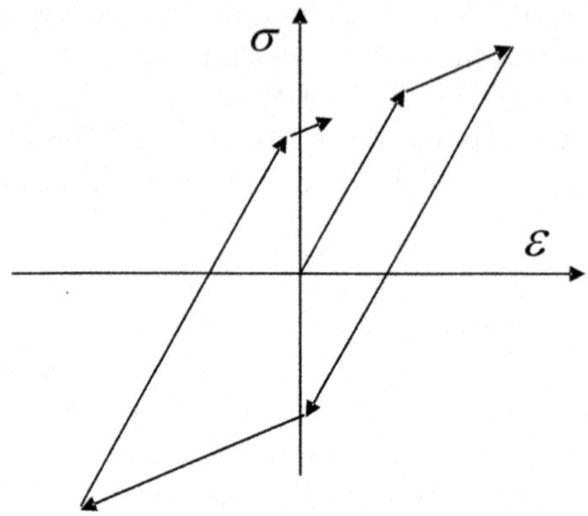

(a) Uniaxial state

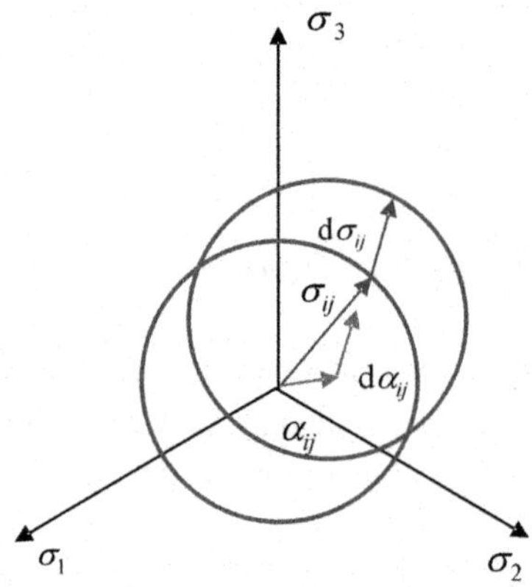

(b) Three-dimension

For the two models, the KH is simulated by a combination of backstresses. The evolution rules of the backstresses are defined as a combination of the Prager model and a relaxation term, and the relaxation term is introduced to simulate the nonlinear evolution of the backstress. Moreover, the evolution rate of the backstress is found to be different at different strain ranges, and the model is modified using a combination of several backstresses with different evolution rules. The evolution for each backstress is defined as follows

$$\mathrm{d}\boldsymbol{\alpha}_\mathbf{i} = C_i \frac{1}{\sigma_{y0}} (\mathbf{S} - \boldsymbol{\alpha}_\mathbf{i}) \mathrm{d}\varepsilon_{\mathrm{eq}}^{p} - \gamma_i \cdot \boldsymbol{\alpha}_\mathbf{i} \cdot \mathrm{d}\varepsilon_{\mathrm{eq}}^{p} \tag{3.7}$$

where C_i and γ_i are model parameters, and the overall backstress can be calculated as

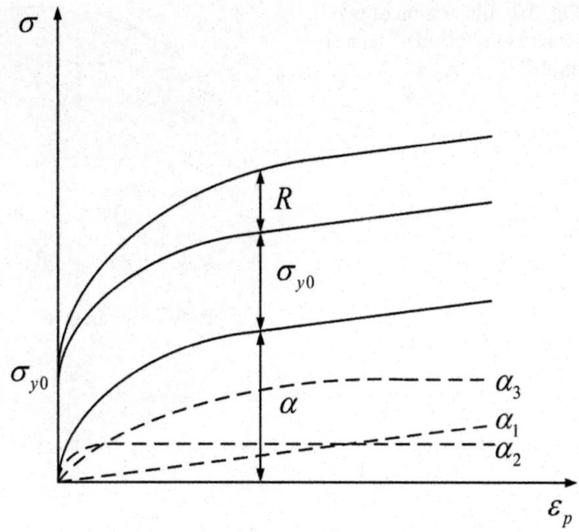

Fig. 3.3 Illustration of Chaboche model with IH under uniaxial stress state

$$\boldsymbol{\alpha} = \sum_{i=1}^{n} \boldsymbol{\alpha}_{\mathbf{i}} \tag{3.8}$$

where n is the number of backstresses. Each backstress has a different evolution law and will cover different strain ranges.

The Chaboche model with IH is a combined hardening model with both KH and IH components, and the formation of the KH component is the same as that of the Chaboche model without IH. The IH rule of the yield surface is given by Zaverl and Lee (1978) as

$$\mathrm{d}R = k(Q_\infty - R)\mathrm{d}\varepsilon_{\mathrm{eq}}^{p} \tag{3.9}$$

where R and $\mathrm{d}R$ are the change and incremental change of the yield surface, respectively, and k is a model parameter to describe the IH rate. Q_∞ is the maximum change of the yield surface. For a uniaxial state, the evolution rule of the IH of the yield surface can be expressed as

$$\sigma = \sigma_{y0} + R = \sigma_{y0} + Q_\infty(1 - \mathbf{e}^{-k \cdot \varepsilon_{eq}^{p}}) \tag{3.10}$$

The combined hardening model under uniaxial stress state is illustrated in Fig. 3.3, where three backstresses are employed, and one of the backstresses is with a linear evolution rule.

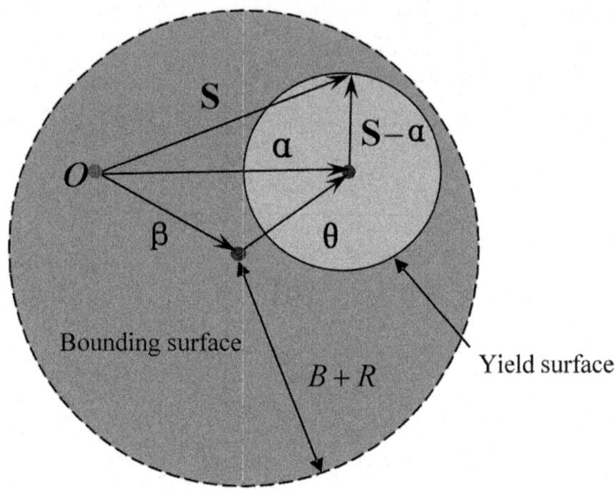

Fig. 3.4 Illustration of two surfaces in Yoshida–Uemori model

3.2.4 Yoshida–Uemori Model

The Yoshida–Uemori model is a two-surface model. This model is a relatively new model, but it has already been incorporated by some FE software including LS-DYNA. It is a KH model with a constant yield surface and a changing bounding surface. The yield surface of the model with Mises yield function can be expressed as

$$f = \sqrt{\frac{3}{2}(\mathbf{S} - \boldsymbol{\alpha}) : (\mathbf{S} - \boldsymbol{\alpha})} - \sigma_{y0} = 0 \tag{3.11}$$

As illustrated in Fig. 3.4, the function of the bounding surface, F, can be expressed as

$$F = \sqrt{\frac{3}{2}(\mathbf{S} - \boldsymbol{\beta}) : (\mathbf{S} - \boldsymbol{\beta})} - (B + R) = 0 \tag{3.12}$$

where $\boldsymbol{\beta}$ is the center of the bounding surface, B and R are initial size and IH component of the bounding surface, respectively.

The relative KH component of the yield surface and the bounding surface is

$$\boldsymbol{\theta} = \boldsymbol{\alpha} - \boldsymbol{\beta} \tag{3.13}$$

Then

$$\mathrm{d}\boldsymbol{\alpha} = \mathrm{d}\boldsymbol{\theta} + \mathrm{d}\boldsymbol{\beta} \tag{3.14}$$

The evolution of the relative KH component between the yield surface and the bounding surface is

$$\mathrm{d}\boldsymbol{\theta} = C\mathrm{d}\varepsilon_{\mathrm{eq}}^{p}\left[\frac{a}{\sigma_{y0}}(\mathbf{S}-\boldsymbol{\alpha}) - \sqrt{\frac{a}{\bar{\theta}}}\boldsymbol{\theta}\right] \tag{3.15}$$

where C is a model parameter, and $\bar{\theta}$ is defined as

$$\bar{\theta} = \sqrt{\frac{3}{2}\boldsymbol{\theta} : \boldsymbol{\theta}} \tag{3.16}$$

and a is defined as

$$a = B + R - \sigma_{y0} \tag{3.17}$$

The evolution rule for $\boldsymbol{\beta}$ is assumed

$$\mathrm{d}\boldsymbol{\beta} = m\left(\frac{2}{3}b\mathrm{d}\boldsymbol{\varepsilon}_{\mathbf{p}} - \boldsymbol{\beta}\mathrm{d}\varepsilon_{\mathrm{eq}}^{p}\right) = m\left(\frac{2}{3}b\frac{3}{2}\frac{\mathbf{S}-\boldsymbol{\alpha}}{\sigma_{y0}}\mathrm{d}\varepsilon_{\mathrm{eq}}^{p} - \boldsymbol{\beta}\mathrm{d}\varepsilon_{\mathrm{eq}}^{p}\right) = m\left(b\frac{\mathbf{S}-\boldsymbol{\alpha}}{\sigma_{y0}} - \boldsymbol{\beta}\right)\mathrm{d}\varepsilon_{\mathrm{eq}}^{p} \tag{3.18}$$

where m is a model parameter.

Substituting Eqs. (3.15) and (3.18) into Eq. (3.14), we can obtain

$$\begin{aligned}\mathrm{d}\boldsymbol{\alpha} = \mathrm{d}\boldsymbol{\theta} + \mathrm{d}\boldsymbol{\beta} &= C\mathrm{d}\varepsilon_{\mathrm{eq}}^{p}\left[\frac{a}{\sigma_{y0}}(\mathbf{S}-\boldsymbol{\alpha}) - \sqrt{\frac{a}{\bar{\theta}}}\boldsymbol{\theta}\right] + m\left(b\frac{\mathbf{S}-\boldsymbol{\alpha}}{\sigma_{y0}} - \boldsymbol{\beta}\right)\mathrm{d}\varepsilon_{\mathrm{eq}}^{p} \\ &= \left[C\frac{a}{\sigma_{y0}}(\mathbf{S}-\boldsymbol{\alpha}) - \sqrt{\frac{a}{\bar{\theta}}}\boldsymbol{\theta} + mb\frac{\mathbf{S}-\boldsymbol{\alpha}}{\sigma_{y0}} - m\boldsymbol{\beta}\right]\mathrm{d}\varepsilon_{\mathrm{eq}}^{p}\end{aligned} \tag{3.19}$$

The IH component of the bounding surface is assumed to be

$$\mathrm{d}R = m(R_{\mathrm{sat}} - R)\mathrm{d}\varepsilon_{\mathrm{eq}}^{p} \tag{3.20}$$

The size of the bounding surface under uniaxial condition is

$$B + R + \overline{\beta} = B + (R_{\mathrm{sat}} + b)\left(1 - \mathrm{e}^{-m\mathrm{d}\varepsilon_{\mathrm{eq}}^{p}}\right) \tag{3.21}$$

where R_{sat} is the saturated value of the bounding surface at infinite plastic strain, and $\overline{\beta}$ is the norm of $\boldsymbol{\beta}$. There is also a non-IH surface which is used to memorize the hardening history. The non-IH surface is commonly called memory surface. The memory surface which can simulate the work hardening stagnation is defined as

$$g_{\sigma} = \sqrt{\frac{3}{2}(\boldsymbol{\beta}-\mathbf{q}) : (\boldsymbol{\beta}-\mathbf{q})} - r = 0 \tag{3.22}$$

where $\mathbf{q}$ and r are the center and radius of the memory surface, respectively.

Motion of the surface g_{σ} is assumed to be

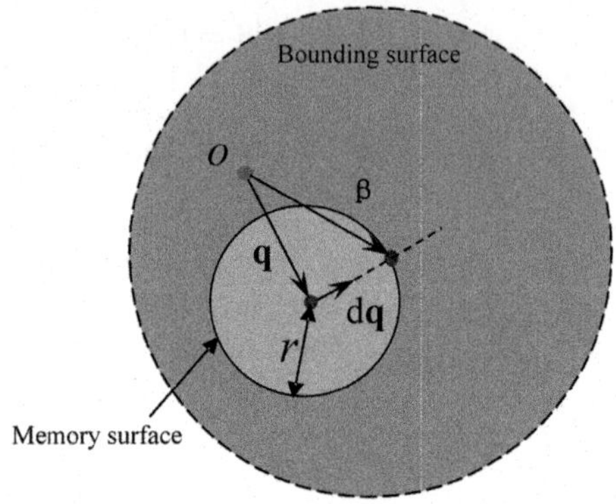

Fig. 3.5 Illustration of memory surface of Yoshida–Uemori model

$$\mathrm{d}\mathbf{q} = \mu(\boldsymbol{\beta} - \mathbf{q}) \tag{3.23}$$

Consistency condition of the surface g_σ leads to

$$\mu = \frac{3(\boldsymbol{\beta} - \mathbf{q}) : \mathrm{d}\boldsymbol{\beta}}{2r^2} - \frac{\mathrm{d}r}{r} \tag{3.24}$$

The evolution of r is assumed to be

$$\mathrm{d}r = h\Gamma, \quad \Gamma = \frac{3(\boldsymbol{\beta} - \mathbf{q}) : \mathrm{d}\boldsymbol{\beta}}{2r} \text{ when } \mathrm{d}R > 0 \tag{3.25}$$

where h is a model parameter which determines the rate of expansion of surface, g_σ.

The memory surface can be illustrated as in Fig. 3.5. To explain the functions of the seven model parameters, the physical meaning of each parameter is explained in Table 3.1, and schematic illustration of the parameters under uniaxial cyclic loading is also given in the literature (Yoshida and Uemori 2002).

3.2.5 Modification of Yoshida–Uemori Model

(a) *Limitations of Yoshida–Uemori Model*
It was reported (Yoshida and Uemori 2002) that the Yoshida–Uemori model was capable of simulating cyclic plasticity of metals at large plastic strain ranges. However, there are several limitations of the model.

(1) It is found that some metals exhibit continuous strain hardening without saturation till fracture of the materials (Jia and Kuwamura 2014a; Shi et al. 2008). However, the IH component of the Yoshida–Uemori model tends to reach a limiting value at large plastic strain, which may lead to considerable error during simulating plasticity of some structural steels under both

Table 3.1 Physical meanings of the model parameters of Yoshida–Uemori model

Parameters	Physical meanings
σ_{y0}	Initial yield stress; A larger value of σ_{y0} commonly gives higher stress amplitudes for both forward and reversed half cycles of the cyclic stress–strain curve
C	Hardening rate of the kinematic hardening component; A larger value of C gives larger curvatures in transient Bauschinger effect zones of the cyclic stress–strain curve
B	Initial size of the bounding surface; A larger value of B gives higher stress amplitudes for both forward and reversed half cycles of the cyclic stress–strain curve, and the increase in the stress amplitudes is independent on strain ranges
R_{sat}	Saturated size of the isotropic hardening component; A larger value of R_{sat} gives higher stress amplitudes for both forward and reversed half cycles of the cyclic stress–strain curve, and the increase in the stress amplitudes is dependent on strain ranges
b	Saturated size of the backstress $\boldsymbol{\beta}$; A larger value of b gives higher stress amplitudes in the forward half cycles, while gives lower stress amplitudes in revered half cycles of the cyclic stress–strain curve
m	Hardening rate of the isotropic hardening component R as well as of the kinematic hardening component $\boldsymbol{\beta}$; A larger value of m gives larger curvatures of the cyclic stress–strain curve
h	Expansion rate of the memory surface; A larger value of h gives lower stress amplitudes for both forward and reversed half cycles of the cyclic stress–strain curve

monotonic and cyclic loadings. The model may underestimate the strain hardening effect in large strain ranges.

(2) The model defines a memory surface in stress space. However, there is a problem with the evolution rule. According to Eq. (3.18), $\boldsymbol{\beta}$ will eventually reach its saturation value under monotonic loading. Then according to Eq. (3.23), $\mathbf{q}$ will eventually become asymptotically equal to $\boldsymbol{\beta}$, since $\boldsymbol{\beta}$ remains fixed. But when $\boldsymbol{\beta} = \mathbf{q}$, then from Eq. (3.22), the memory surface, g_σ, has zero size of r, which will make the evolution condition indeterminate. This will lead to the fact that the IH of the bounding surface is always active when $\boldsymbol{\beta}$ is saturated and the memory surface will become useless. For example, when $\boldsymbol{\beta}$ reached a limiting value in large strain ranges, the IH component will always increase, and the stress under constant-amplitude loading will be overestimated due to this problem.

(3) The Yoshida–Uemori model cannot simulate the yield plateau. For the evolution rules of the backstresses are coupled with each other, it may be difficult for the model to simulate the structural steels with yield plateau. If the yield plateau is not well simulated in the model, the calibrated model parameter C may not give an accurate result for mild steel from a tension coupon test, which will also affect the values of the other parameters. How-

ever, the Chaboche models don't have this problem, since the backstresses are independent from each other.

(b) *Modification of Isotropic Hardening Component*
The Yoshida–Uemori model assumes the same evolution rule with the one proposed by Zaverl and Lee (1978) for the IH component of the bounding surface. The size of the bounding surface of the Yoshida–Uemori model under a uniaxial state can be given as

$$B + R = B + R_{\mathrm{sat}}(1 - \mathrm{e}^{-m\mathrm{d}\varepsilon_{\mathrm{eq}}^{p}}) \tag{3.26}$$

It can be found that the IH component, R, tends to saturate to a constant value, R_{sat}. However, it is found in Chap. 2 that the true stress at large plastic strain still increases approximately linearly with the increase of plastic strain without saturation (Jia and Kuwamura 2014b). Therefore, a linear hardening function is superimposed with the one proposed by Zaverl and Lee (1978) in the modified Yoshida–Uemori model as

$$\mathrm{d}R = mR_{\mathrm{sat}}\mathrm{e}^{-m\varepsilon_{\mathrm{eq}}^{p}}\mathrm{d}\varepsilon_{\mathrm{eq}}^{p} + m_l\mathrm{d}\varepsilon_{\mathrm{eq}}^{p} \tag{3.27}$$

where m and m_l are model parameters. The hardening rate at large plastic strain is decided by the hardening modulus in the linear term, m_l, while the hardening rate at small plastic strain ranges, e.g., during the pre-necking stage, is determined by both m and m_l.

The expression of R under a uniaxial state can be given as

$$R = R_{\mathrm{sat}}(1 - \mathbf{e}^{-m\mathrm{d}\varepsilon_{\mathrm{eq}}^{p}}) + m_l\mathrm{d}\varepsilon_{\mathrm{eq}}^{p} \tag{3.28}$$

It can be found from Eq. (3.28) that the first term of the IH component of the bounding surface adopted in the present model will saturate at large plastic strain, while the second term will still evolve linearly at large plastic strain, which is in accordance with the classical experimental results of metals under high hydrostatic pressure to fracture carried out by Bridgman (1952).

(c) *Modification of Memory Surface*
The memory surface generalized by Ohno (1982) in strain space as illustrated in Fig. 3.6 is adopted herein, which can always ensure the determination of the evolution condition for the IH of the bounding surface. The expression of the memory surface in strain space is defined as

$$g_{\varepsilon} = \sqrt{\frac{2}{3}(\boldsymbol{\varepsilon}_{\mathbf{p}} - \mathbf{q}):(\boldsymbol{\varepsilon}_{\mathbf{p}} - \mathbf{q})} - r = 0 \tag{3.29}$$

where $\mathbf{q}$ and r are the center and radius of the memory surface, respectively.

Similarly, evolution of the memory surface in strain space, g_{ε}, is defined as

$$\mathrm{d}\mathbf{q} = \mu(\boldsymbol{\varepsilon}_{\mathbf{p}} - \mathbf{q}) \tag{3.30}$$

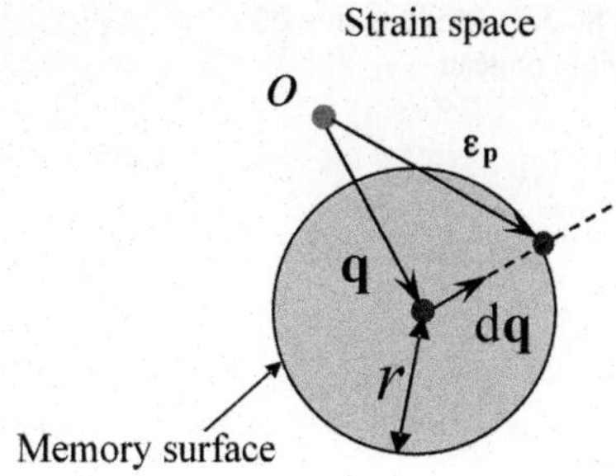

Fig. 3.6 Illustration of memory surface proposed by Ohno

where the initial value of $\mathbf{q}$ is zero.

The consistency condition of the surface, g_ε, leads to

$$\frac{\partial g_\varepsilon}{\partial \boldsymbol{\varepsilon}_\mathrm{p}} : \mathrm{d}\boldsymbol{\varepsilon}_\mathrm{p} + \frac{\partial g_\varepsilon}{\partial \mathbf{q}} : \mathrm{d}\mathbf{q} + \frac{\partial g_\varepsilon}{\partial r} \cdot \mathrm{d}r = 0 \tag{3.31}$$

Then

$$\mu = \frac{2(\boldsymbol{\varepsilon}_\mathrm{p} - \mathbf{q}) : \mathrm{d}\boldsymbol{\varepsilon}_\mathrm{p}}{3r^2} - \frac{\mathrm{d}r}{r} \tag{3.32}$$

The evolution rule of r is assumed to be

$$\mathrm{d}r = h\eta, \quad \eta = \frac{2(\boldsymbol{\varepsilon}_\mathrm{p} - \mathbf{q}) : \mathrm{d}\boldsymbol{\varepsilon}_\mathrm{p}}{3r} \text{ when } \mathrm{d}R > 0 \tag{3.33}$$

where h is a model parameter which determines the rate of expansion of the surface, g_ε, and r will not change if dR is less than zero. According to Eq. (3.33), there will be no IH if the current strain state is inside the memory surface or the angle between the current incremental plastic strain tensor and the normal of the memory surface is larger than 90°. With this rule, the stress stabilization of metals under constant-amplitude loading can be simulated, while the stress predicted by plasticity models without a memory surface will continue increasing without saturation under constant-amplitude loading.

(d) *Modification of Yield Plateau*
The cyclic plasticity within the yield plateau is very complicated, and an elaborate simulation of the yield plateau has been given by Mahan et al. (2011). A simplified proposal is presented by Yoshida and Uemori (2002) to simulate the yield plateau, while it is defined in stress space, and is difficult to define the model parameter related to the size of yield plateau. Herein, it is assumed that no hardening takes place when the equivalent plastic strain is less than the size of yield plateau denoted by a scalar, $\varepsilon_{\text{plateau}}$, as illustrated in Fig. 3.7. This condition is defined as

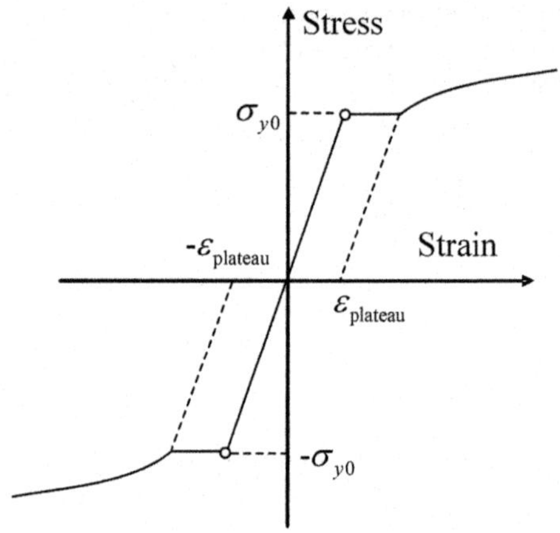

Fig. 3.7 Modification of yield plateau

$$C = 0, \quad m = 0 \text{ when } \varepsilon_{eq}^{p} < \varepsilon_{plateau} \tag{3.34}$$

3.3 Experiment

3.3.1 Material

The material of the coupons and specimens is SS400 (JIS G3101 2015), and the material properties and chemical composition provided are the same as those listed in Tables 2.1 and 2.2, respectively.

3.3.2 Sampling of Specimens

All the specimens were taken from a 65 mm thick steel plate, which is the same as the tension coupons in Chap. 2. As is known, the impurity substance usually accumulates near the neutral surface of thick steel plates, so all the specimens were taken from the part near the plate surface as shown in Fig. 3.8 to avoid the influence. Seven hourglass-type specimens, two ordinary-type tensile coupons with a parallel part, and additionally a straight round bar were manufactured. The tension coupons are those introduced in Chap. 2, and herein employed again to determine the values of plasticity model parameters. The straight round bar, which has a larger sectional area than the hourglass-type specimen, was used to pick up mechanical slips generated within the testing machine and the setup during cyclic loadings.

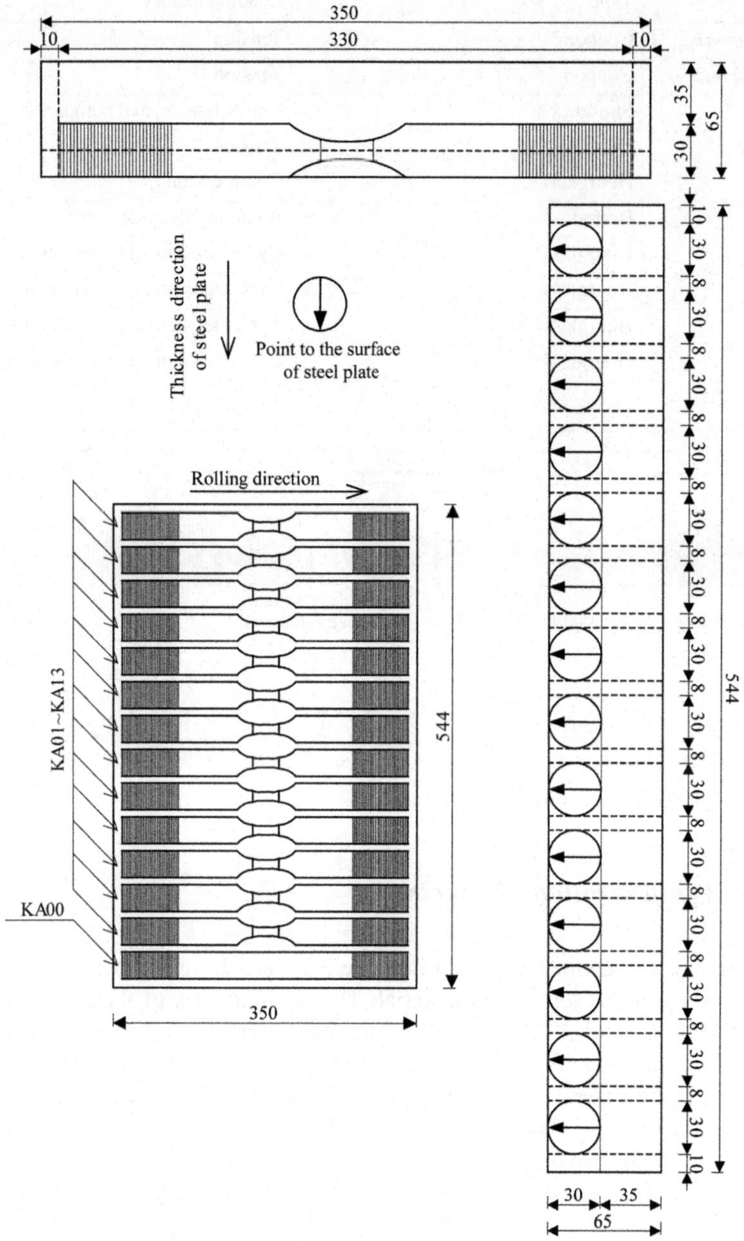

Fig. 3.8 Sampling of specimens

Table 3.2 Specimens of cyclic tension and compression test of SS400

No.	Type	Loading history
KAcoupon1	Coupon	Tension
KAcoupon2	Coupon	Tension
KA00	Straight bar	Cyclic tension and compression
KA01	Hourglass	Tension
KA02	Hourglass	Compression
KA03	Hourglass	A single full cycle
KA04	Hourglass	Cyclic tension and compression
KA05	Hourglass	Cyclic tension and compression
KA06	Hourglass	Cyclic tension and compression
KA07	Hourglass	Cyclic tension and compression

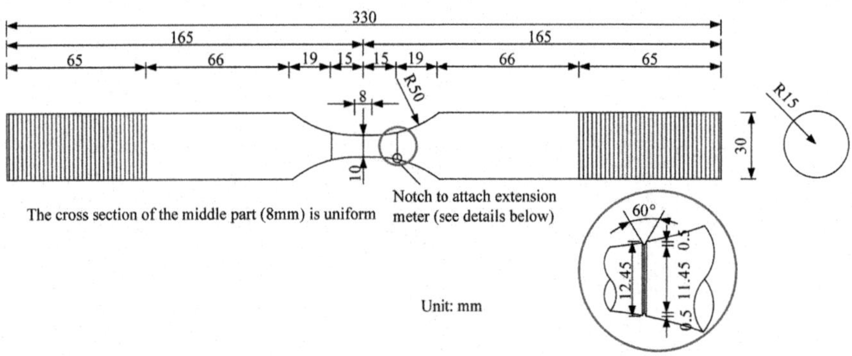

Fig. 3.9 Configuration of hourglass-type specimens

3.3.3 Configuration of Specimens

Hourglass-type specimens as listed in Table 3.2 were designed to carry out cyclic tension-compression test of the material. The configuration of the hourglass-type specimens is as shown in Fig. 3.9. A minimum diameter of 10 mm was decided considering the capacity of the machine, and also a parallel part was designed to make sure that the stress across the cross section is uniform in moderate strain range. However, the length of the parallel part could not be too long since it may buckle under compression. A length of 8 mm was decided considering the above two factors. Two small notches were designed to attach the elongation meter during the test, and the transition parts were designed to arc-shaped to prevent the specimens from premature instability.

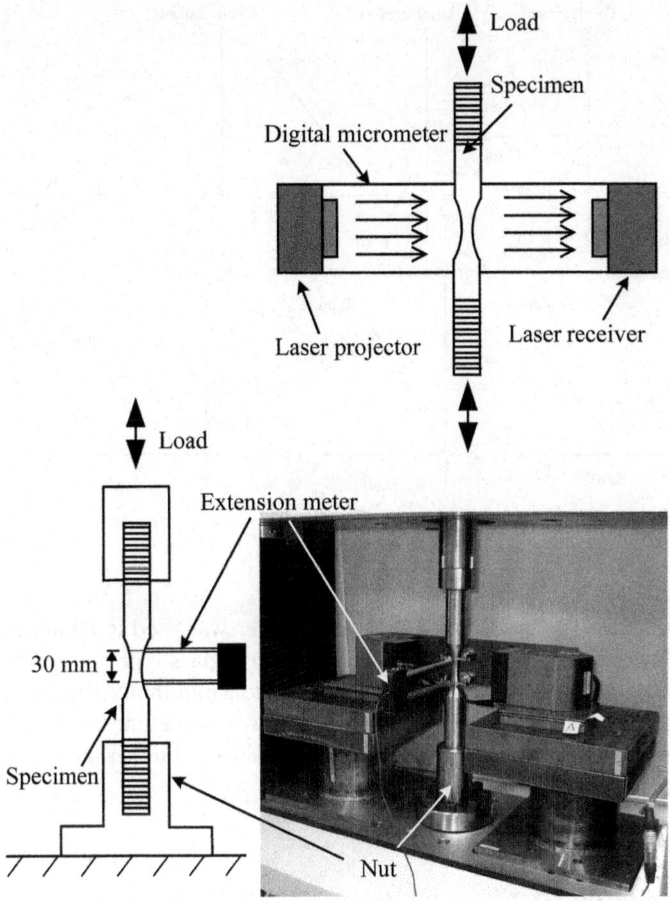

Fig. 3.10 Test setup of hourglass-type specimens

3.3.4 Test Setup

The tests on monotonic tension coupons are the same as those in Chap. 2 and have been discussed. For the tests of the hourglass-type specimens, the test setup was designed to prevent buckling of the specimens using two nuts attached to the two ends of the specimen as shown in Fig. 3.10. The test machine of AG-50KNX+ 250 mm made by SHIMAZU Cooperation with a load capacity of ±50,000 N and a displacement capacity of 250 mm was used to carry out the tests. For all the tests, two-dimensional digital micrometers (Two types of heads are used, i.e., TM-040 and TM-065, the controller is of the same series, TM-3000) were adopted to measure the minimum diameters of the parallel parts of the specimens, from the data of which the true strain could be obtained.

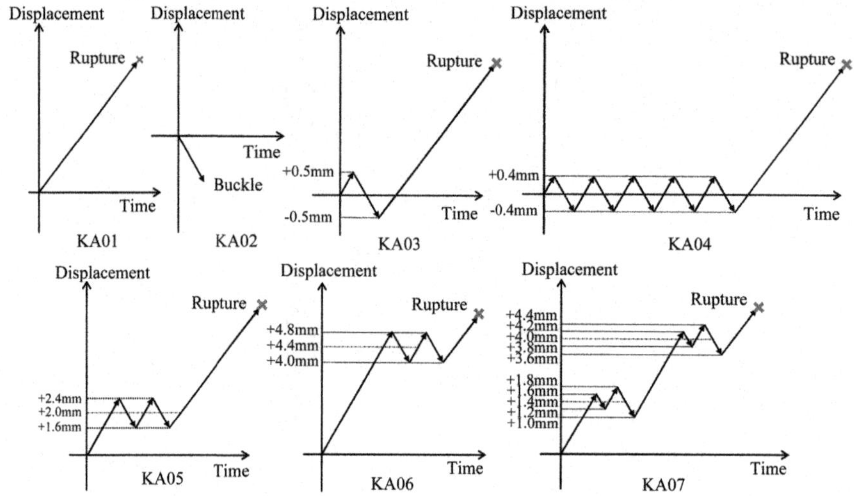

Fig. 3.11 Loading histories of specimens

For hourglass-type specimens, the extensometer was used to obtain the relative change of length within the extensometer. The hourglass-type specimens were all under displacement control by the extensometer to avoid minor slip within the test system. However, since the value of extensometer was sometimes too little to control the loading rate, the stroke of the test machine was used to set the loading rate to achieve a stable loading process.

3.3.5 Loading Histories for Hourglass-Type Specimens

Tests on seven hourglass-type specimens were carried out under various loading histories as illustrated in Fig. 3.11. The specimens *KA01* and *KA02* were loaded under monotonic tension and compression, respectively, and *KA03* was tested under a single full cycle, and the specimen *KA04* was tested under five-constant-amplitude-cycle and then pulled to fracture. The specimens *KA05* and *KA06* were loaded under two-constant-amplitude-cycle before and after necking initiation of the specimens, respectively, and then pulled to fracture. The specimen *KA07* was loaded cyclically both before and after necking initiation, and then pulled to fracture.

Table 3.3 Model parameters of three hardening models calibrated using true stress obtained by modified weighted average method

Prager model		Chaboche models			Yoshida–Uemori model		
			Without isotropic hardening	With isotropic hardening		Original	Modified
σ_{y0}	255.9	σ_{y0}	255.9	255.9	σ_{y0}	255.9	255.9
C_0	1429.0	C_1	97.2	26.9	C	338.7	332.8
		C_2	97.2	26.9	B	277.3	321.7
		C_3	3763.0	1617.2	R_{sat}	196.0	137.7
		γ_1	0	0	b	194.9	82.9
		γ_2	0	0	m	7.9	18.1
		γ_3	13.7	10.7	h	0.5	0.5
		k	/	5.8	m_l	/	236.2
		Q_∞	/	227.8	$\varepsilon_{plateau}$	/	0.0148

Note Unit of σ_{y0}, C_0, C_1, C_2, C_3, Q_∞, C, B, R_{sat}, b, and m_l: MPa
Unit of γ_1, γ_2, γ_3 and k, m, h, and $\varepsilon_{plateau}$: dimensionless

3.4 Simulation

3.4.1 Calibration of Model Parameters Using Only Tension Coupon Test

(a) *Calibration Method of Model Parameters of Prager Model*
The Prager model has only two model parameters including the yield stress and the KH modulus C_0 in Eq. (3.6). The yield stress can be obtained straightforwardly from the coupon test result, and C_0 should be given depending on the strain range that the problem is concerned with. For example, it is often given by the following equation, the result of which is as shown in Table 3.3.

$$C_0 = \frac{\sigma - \sigma_{y0}}{\varepsilon_{eq}^p} \tag{3.35}$$

where σ is the yield stress at a finite equivalent plastic strain ε_{eq}^p. Herein, the initial yield stress σ_{y0} is 255.9 N/mm^2 obtained from the test result. Since this study focuses on plasticity at large plastic strain ranges till fracture, the true stress before fracture of *KAcoupon2* is adopted for σ, which is equal to 837.5 N/mm^2 at a plastic strain of 0.90.

(b) *Calibration Method of Model Parameters of Chaboche Models*
For the Chaboche model without IH, the model parameters that must be determined are the initial yield stress, σ_{y0}, and hardening parameters, C_i and γ_i in

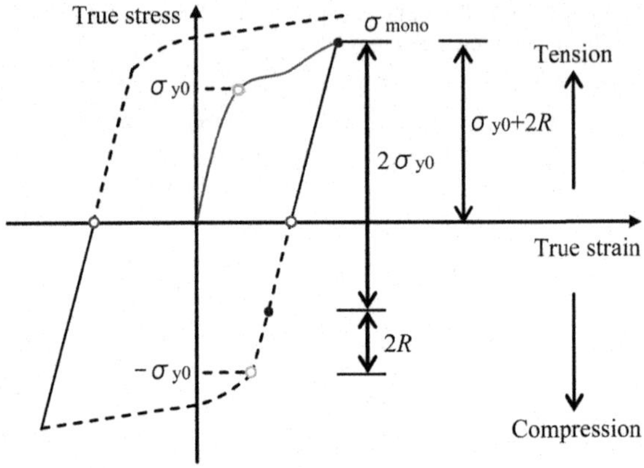

Fig. 3.12 Calibration of model parameters for isotropic hardening component of Chaboche model with IH

Eq. (3.7). It is straightforward to calibrate the parameters using the monotonic tension coupon test results. First, the true stress–true strain data till fracture of a coupon can be obtained using the modified weighted average method proposed in Chap. 2. Then the parameters, C_i and γ_i, can be calibrated by inputting the yield stress and equivalent plastic strain in tabular form in ABAQUS. The model parameters will be printed out in a corresponding result file.

However, for the Chaboche model with IH, the IH parameters that have to be determined are the maximum change in the size of the yield surface, Q_∞, and the rate of the change in the size of the yield surface, k in Eq. (3.9). The IH parameters have to be determined simultaneously with the KH component. The input data related to the calibration of the KH component, i.e., C_i and γ_i, is the same as that of the Chaoche model without IH. Herein, a method to determine the input data to calibrate the parameters related to the IH component from only tension coupon test results is required. It is found in the cyclic tests carried out on hourglass-type specimens that the yield stress under compression in every cycle is close to the initial yield stress σ_{y0} as illustrated in Fig. 3.12. Under the assumption that these values are equal, the size of the subsequent yield surface of the Chaboche model with IH, $\sigma_{y0} + R$, can be obtained using the monotonic coupon test result

$$\sigma_{\text{mono}} = \sigma_{y0} + 2R \tag{3.36}$$

where σ_{mono} is the post-yield uniaxial true stress obtained by the modified weighted average method proposed in Chap. 2. This assumption is found to be the same as the principle previously proposed by Kuhlmann-Wilsdorf and Laird (1979) based on the Cottrell's treatment (Cottrell 1953) of hysteresis loops, and this rule has also been found reliable for copper single crystals. Then the model parameters, Q_∞ and k, related to the IH component can be calibrated through the input data in tabular form (post-yield stress, equivalent plastic strain) in ABAQUS. The optimal model parameters are listed in Table 3.3. Three backstresses are commonly necessary for the

Chaboche models (ABAQUS 2010). The calibrated parameters of the two Chaboche models, γ_1 and γ_2, are zero, which indicates that two backstresses are with linear evolution rules.

Special attention should be paid to the input uniaxial true stress–true strain for calibration of the model parameters. For example, the input maximum true strain should be larger than the maximum equivalent strain reached in numerical simulations. The number and selection of data points are also important for the calibration result. For example, the total number of data points should not be fewer than the number of model parameters to ensure the calibrate result be the unique combination of parameters. Moreover, the critical data points, e.g., the initial yield point, the end point of the yield plateau, should be included in the input. More data points within the strain range before necking should be incorporated since the stress–strain curve is more curved in the range.

(c) *Calibration Method of Model Parameters of Original and Modified Yoshida–Uemori Models*
The model parameters for the original Yoshida–Uemori model which have to be calibrated are the initial yield stress, σ_{y0}, and the parameters related to the yield surface, the bounding surface, and the memory surface, C, m, B, b, $R_{sat,}$ and h. σ_{y0} can be obtained directly from the monotonic coupon test in the same way as the other models. An optimization analysis was carried out to obtain the optimized combination of the model parameters (C, m, B, b, R_{sat}, h) using a script written by the authors. The objective function of the optimization analysis is defined as $\sqrt{\sum(\sigma_{FE} - \sigma_{test})^2/\sigma_{test}^2}$, where σ_{FE} and σ_{test} are the true stresses obtained from the numerical and experimental results of the monotonic coupon tests, respectively. According to the evolution rules employed by Chaboche and Dang (1979), h can be specified as 0.5 in the optimization analysis.

Likewise, calibration of the model parameters of the modified Yoshida–Uemori model is similar to that of the original Yoshida–Uemori model using only tension coupon test results. The only difference between the calibration processes of the two models is the number of variables in the objective function during the optimization analysis. The objective function for the modified Yoshida–Uemori model has two additional variables, i.e., m_l and $\varepsilon_{plateau}$.

3.4.2 FE Modeling

Numerical simulations were carried out to simulate the force–deformation of the specimens with axisymmetric half models using CAX8 elements in ABAQUS (2010). Symmetric boundary conditions were applied to the symmetric edges, and the rupture displacement of the corresponding tests was enforced at the top edge. The whole model and partial model as shown in Fig. 3.13 were simulated. The tests of hourglass-type specimens were displacement controlled by the extensometer with a gage length of 30 mm. Since the simulation results of the whole model and the partial model give

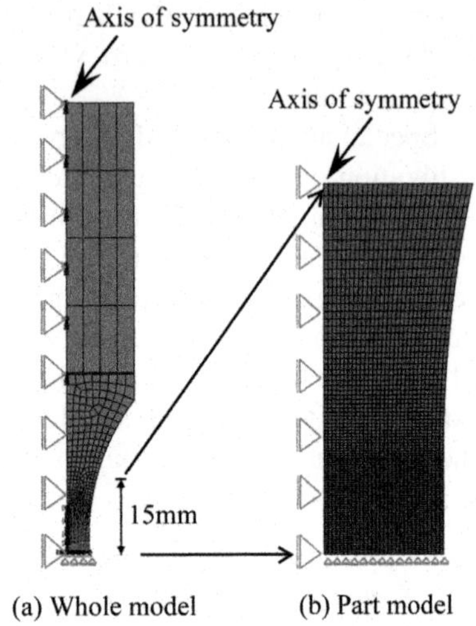

Fig. 3.13 Whole and partial FE models

almost identical force–deformation curve for a monotonic tension test, herein only the part within the extensometer was simulated with the partial model. Numerical simulations were carried out using the three selected plasticity models and the modified Yoshida–Uemori model. The calibration methods for the model parameters of each model have been given in the preceding sections. The corresponding model parameters of each model are listed in Table 3.3.

3.5 Comparison of Experiment and Simulation

The force–deformation curves of *KA01*, *KA03–KA07* are shown in Fig. 3.14, and the results of *KA02* are not shown due to premature buckling before fracture.

Based on the comparison between the test results and the corresponding FE ones, it is found that the Prager model is too simple to simulate the cyclic behaviors of metal plasticity at large plastic strain. The test result of *KA03* shows that the Chaboche model without IH underestimates the load in subsequent loading cycles. The Chaboche model with IH gives much better simulation results of *KA03*, while the stress of the Chaboche model with IH doesn't tend to stabilize under constant-amplitude loading (*KA04*).

The original Yoshida–Uemori model cannot simulate the tests well (*KA01*, *KA03*) due to the aforementioned three limitations of the model. However, the modified Yoshida–Uemori model proposed in this chapter gives an acceptable accuracy for the behaviors of mild steel under both monotonic tension (*KA01*), and complicated cyclic loading histories (*KA03–KA07*). The test result of *KA06* indicates that loading cycles close to fracture strain can result in fast material deterioration, and the effect of micro-voids may have some effects on the cyclic behaviors of the material.

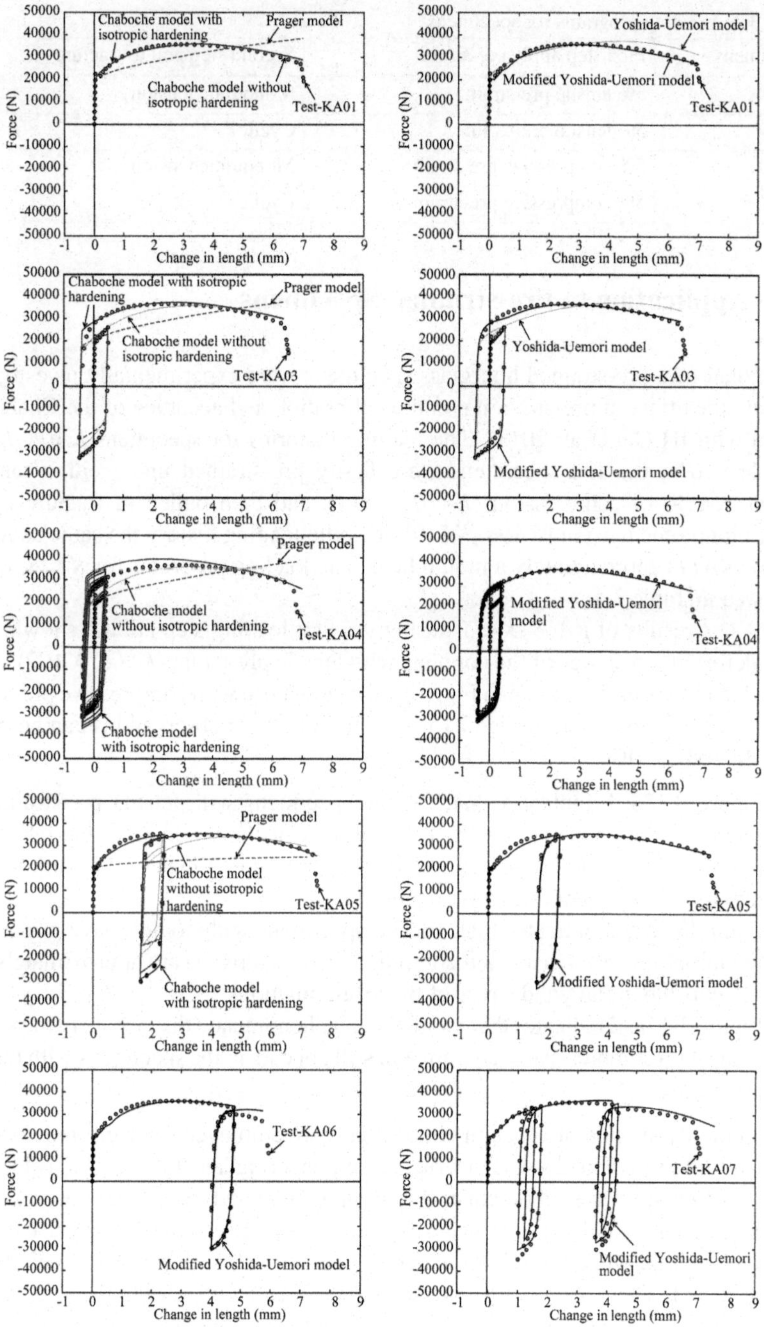

Fig. 3.14 Comparison between test and numerical results

Table 3.4 Loading programs for specimens

Specimens	First step apply pre-strain	Second step pull to fracture
KA08	5% tensile pre-strain	Monotonic tension
KA09	5% tensile pre-strain	Cyclic
KA10	5% compressive pre-strain	Monotonic tension
KA11	5% compressive pre-strain	Cyclic

3.6 Application to Pre-strained Specimens

Additional four pre-strained hourglass specimens were experimentally investigated to study the effect of pre-straining on the calibration and accuracy of the Chaboche model with IH (Jia et al. 2013). The loading histories for specimens *KA08*–*KA11* include two steps. The specimens were firstly pre-strained under either tension (*KA08*, *KA09*) or compression (*KA10*, *KA11*), and then pulled to fracture under either monotonic tension (*KA08*, *KA10*) or cyclic loading history the same as *KA07* (*KA09*, *KA11*) approximately a month later. The loading programs for *KA08*–*KA11* are listed in Table 3.4.

The test results of *KA08*–*KA11* during the first loading step are not shown. The load–deformation curves of the coupons with tensile pre-strain (*KA08*, *KA09*) at the second loading step are shown in Fig. 3.15a, b, and those with compressive pre-strain (*KA10*, *KA11*) are shown in Fig. 3.15c, d. The following observations can be made from the test results.

(1) Figure 3.15a, b indicates that, for the specimens with tensile pre-strain, the initial tensile yield stress is higher than that of the virgin material, while the initial reversal compressive yield stresses are approximately the same as the initial yield stress of the virgin material.
(2) Figure 3.15c, d indicates that, for the specimens with compressive pre-strain, the initial reversal tensile and compressive yield stresses are approximately the same as the initial yield stress of the virgin material.
(3) Figure 3.15a–d indicates that yield plateau phenomenon disappears in the specimens with compressive pre-strain, but still appears in the specimens with tensile pre-strain.

Commonly, monotonic tension coupon tests are employed to determine material properties of pre-strained steels in structural engineering, and the initial tensile yield stress is of great interest in practice. Based on the first and the second observations, the initial tensile yield stress of material with tensile pre-strain will increase due to strain hardening, while that of the material with compressive pre-strain will almost be unaffected. The reason is that Bauschinger effect occurs during the tension test of the materials with compressive pre-strain, and offset the increased stress caused by the strain hardening.

The main purpose is to accurately simulate the load–deformation curves of *KA09* and *KA11* at the second loading steps. Two approaches are taken in the simulation.

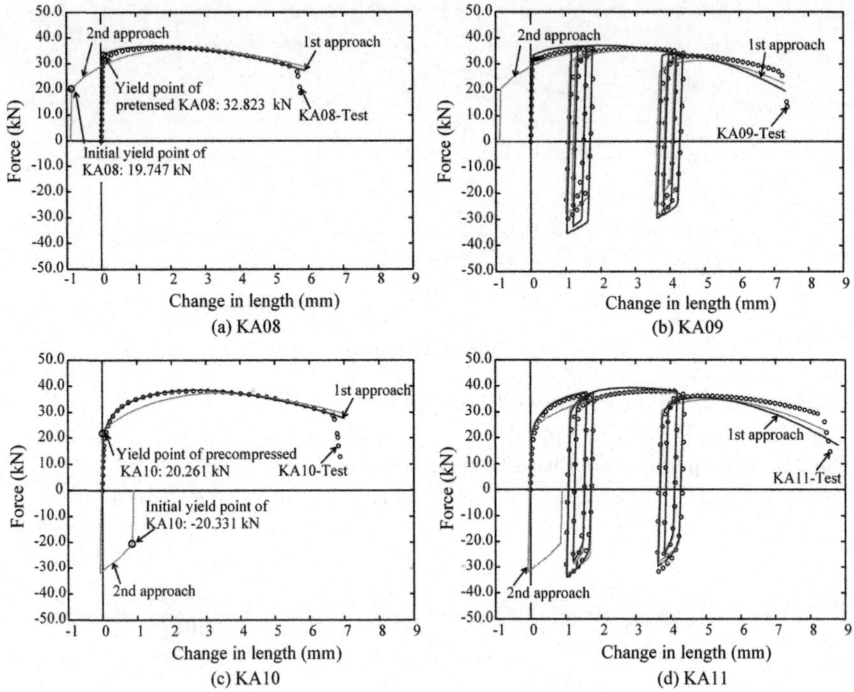

Fig. 3.15 Comparison of experimental and numerical results for hourglass specimens

The first approach uses plasticity model parameters calibrated from tension coupon test results of the pre-strained materials, while the second approach uses those calibrated from the tension coupon test result of the original material along with the pre-strain histories of the materials. In practice, the first approach is more commonly employed, since the coupon test results of virgin materials as well as the pre-strain histories are often unknown.

Likewise, numerical simulations were conducted to investigate the capacity of the Chaboche model with IH for cyclic plasticity of pre-strained structural steel. For the first calibration approach, the load–change in diameter curve of the hourglass specimen, *KA01*, is almost the same as that of the coupon test as shown in Fig. 3.16. Therefore, the test results of *KA08* and *KA10* at the second loading step can be deemed as the monotonic tension coupon test results of the pre-strained materials. For the second approach, the aforementioned coupon test results were used. The model parameters of the Chaboche model using the two approaches were listed in Table 3.5. The γ_1 of the virgin material and the pre-strained materials are equal to zero, indicating that there is a backstress with a linear evolution rule for each material. The calibrated yield stress of the material with tensile pre-strain is much higher than that of the virgin material, while the calibrated yield stress of the material with compressive pre-strain has a value close to that of the virgin material.

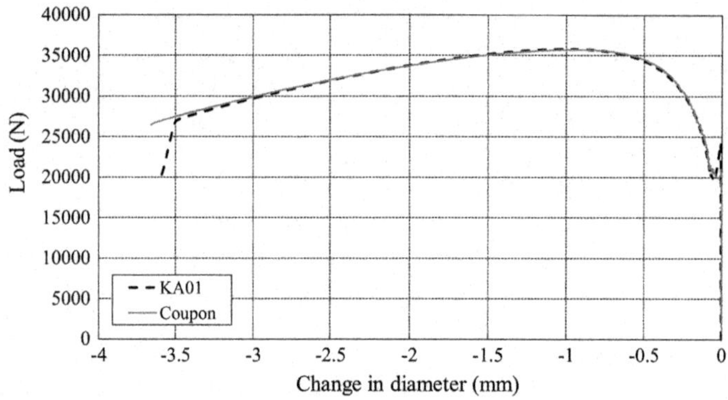

Fig. 3.16 Comparison of coupon test result and *KA01*

Table 3.5 Model parameters of Chaboche model with IH

Model parameters	Virgin material	*KA08* with tensile pre-strain (second loading step)	*KA10* with compressive pre-strain (second loading step)
σ_{y0}	255.9	446.3	246.5
C_1	26.9	135.8	104.5
C_2	26.9	301.7	1267.3
C_3	1617.2	286.7	3992.8
γ_1	0	0	0
γ_2	0	12.7	13.4
γ_3	10.7	13.5	84.4
k	5.8	2. 89	10.97
Q_∞	227.8	162.2	197.0

Note Unit of σ_{y0}, C_1, C_2, C_3, Q_∞: MPa.
Unit of γ_1, γ_2, γ_3 and k: dimensionless

The experimental results at the second loading step and the corresponding numerical results are compared in Fig. 3.15. Both the first and the second approach can well simulate the monotonic tension tests of the materials with tensile and compressive pre-strain well, as shown in Fig. 3.15a, c. The first approach slightly overestimates the compressive stress of the steel with tensile pre-strain as shown in Fig. 3.15b, while accurately predicts the forces for the steel with compressive pre-strain, as shown in Fig. 3.15d. The tensile pre-strain results in an increase in the initial tensile yield force due to strain hardening, while the Bauschinger effect offsets the increase in the yield force resulted by the compressive pre-strain. The initial tensile yield force of the material with compressive pre-strain is thus close to that of the virgin material. Therefore, the calibration method of the Chaboche model gives favorable evaluation of the cyclic plasticity of materials with compressive pre-strain using tension coupon test results, while overestimates the IH of the materials with tensile pre-strain. For the second approach, the numerical results compare well with the experimental results for both materials with tensile and compressive pre-strain.

3.7 Summaries

Plasticity models employed in FE analysis were investigated in comparison with experimental cyclic tests in order to find the most promising model to simulate the cyclic behaviors till fracture of mild steel, from which the following conclusions can be drawn

(1) Prager model cannot simulate the monotonic or cyclic behaviors of mild steel well.
(2) The Chaboche model without isotropic hardening (IH) can simulate the monotonic test results of mild steel well, but it underestimates the load in reversal cycles for cyclic loading.
(3) The Chaboche model with IH can predict both the monotonic and cyclic behaviors of mild steel well, though the stress under cyclic loading with constant strain amplitudes does not tend to stabilize. This results in a slight overestimation of the stress when the loading history of constant strain amplitudes is concerned.
(4) The Yoshida–Uemori model with the model parameters calibrated from monotonic tension coupon tests cannot simulate the monotonic or cyclic behavior of mild steel well.
(5) The modified Yoshida–Uemori model with the modifications of post-necking hardening, memory surface, and yield plateau proposed in this study can predict both the monotonic and cyclic behaviors of mild steel using only the monotonic tension test results with good accuracy.
(6) The calibration method for the parameters of the Chaboche model with IH using the virgin material is found to be more accurate if the pre-strain histories are known. The calibration method with the pre-strained tension coupon can also predict the cyclic plasticity of structural steels with acceptable accuracy.

References

ABAQUS (2010) ABAQUS standard manual (Version 6.10). Karlsson & Sorensen Inc, Hibbitt. Pawtucket (RI, USA)

Bridgman PW (1952) Studies in large plastic flow and fracture. McGraw-Hill, New York

Chaboche J-L, DANG V (1979) Modelization of the strain memory effect on the cyclic hardening of 316 stainless steel. ONERA, TP no 1979-109

Chaboche JL (1986) Time-independent constitutive theories for cyclic plasticity. Int J Plast 2:149

Cottrell AH (1953) Dislocations and plastic flow in crystals. Oxford University Press, London

Dafalias YF, Popov EP (1975) A model of nonlinearly hardening materials for complex loading. Acta Mech 21:173–192

Dafalias YF, Popov EP (1976) Plastic internal variables formalism of cyclic plasticity. J Appl Mech 43:645–651

Frederick CO, Armstrong PJ (2007) A mathematical representation of the multiaxial Bauschinger effect. Mater High Temp 24:1–26

Iwan WD (1967) On a class of models for the yielding behavior of continuous and composite systems. J Appl Mech 34:612–617

Jia L-J, Kuwamura H (2014a) Prediction of cyclic behaviors of mild steel at large plastic strain using coupon test results. J Struct Eng (ASCE) 2(140):04013056

Jia L-J, Kuwamura H (2014b) Ductile fracture simulation of structural steels under monotonic tension. J Struct Eng (ASCE) 5(140):04013115

Jia L-J et al (2013) Prediction of cyclic large plasticity for prestrained structural steel using only tensile coupon tests. Front Struct Civ Eng 4(7):466–476

JIS G3101 (2015) Rolled steels for general structure. Japanese Industrial Standards Committee, Tokyo

Krieg RD (1975) A practical two surface plasticity theory. J Appl Mech 42:641–646

Kuhlmann-Wilsdorf D, Laird C (1979) Dislocation behavior in fatigue II. Friction stress and back stress as inferred from an analysis of hysteresis loops. Mater Sci Eng 37:111–120

Mahan M et al (2011) SANISTEEL: simple anisotropic steel plasticity model. J Struct Eng (ASCE) 137:185–194

Mamaghani IHP et al (1995) Cyclic behavior of structural steels. I: Experiments. J Eng Mech (ASCE) 11(121):1158–1164

Mróz Z (1967) On the description of anisotropic workhardening. J Mech Phys Solids 15:163–175

Ohno N (1982) A constitutive model of cyclic plasticity with a nonhardening strain region. J Appl Mech 49:721–727

Ohno N, Kachi Y (1986) A constitutive model of cyclic plasticity for nonlinear hardening materials. J Appl Mech 53:395–403

Prager W (1949) Recent developments in the mathematical theory of plasticity. J Appl Phys 20:235–241

Shi MF et al (2008) Determination of nonlinear isotropic/kinematic hardening constitutive parameters for AHSS using tension and compression tests. Numisheet 2008Interlaken, Switzerland, 137–142

Shen C et al (1995) Cyclic behavior of structural steels. II: Theory. J Eng Mech (ASCE) 11(121):1165–1172

Yoshida F, Uemori T (2002) A model of large-strain cyclic plasticity describing the Bauschinger effect and workhardening stagnation. Int J Plast 18:661–686

Zaverl F Jr, Lee D (1978) Constitutive relations for nuclear reactor core materials. J Nucl Mater 75:14–19

Chapter 4
Ductile Crack Initiation of Structural Steel under Monotonic Loading

Abstract In this chapter, mesoscopic ductile fracture models based on the void growth theory are briefly reviewed. A void growth fracture model in incremental form is derived based on the Rice–Tracey theory and the Miner's rule, where the emphasis is the crack initiation rule and crack propagation is assumed to occur once a crack initiates. A method to calibrate the fracture model parameter using tension coupon test results is described. The ductile fracture model mainly for crack initiation under monotonic loading is validated by comparison with several experimental results of notched tension coupon test results made of different structural steels, where different stress triaxialities are produced through different configurations of the notches. Comparison results indicate that the void growth ductile fracture model can well predict crack initiation of structural steels under high triaxialities.

4.1 Introduction

4.1.1 Background

Both brittle and ductile fracture of structural steel members has been observed in laboratory tests and in actual buildings damaged in strong earthquakes, e.g., the 1994 Northridge earthquake (Mahin 1998) and the 1995 Kobe earthquake (AIJ 1995). It has been found that brittle fracture of steel structures is triggered by a ductile crack initiated at a notched surface after undergoing noticeable plastic strain (Kuwamura and Yamamoto 1997). Many structures have been reported to crack at beam-to-column connections during the two strong earthquakes. Cracks of bracings and columns initiating after the ULCF loading have also been observed in the Kobe earthquake (AIJ 1995; Kuwamura Lab 1995). A lot of research work related to prevention of brittle failure of steel structural members has been carried out (e.g., Kuwamura 1998, 2003; Kuwamura and Akiyama 1994; Kuwamura et al. 2003). An interim guideline to prevent brittle failure of beam-to-column connections of steel frames has been published (The Building Center of Japan 2003) in Japan. However, research on ductile fracture is inactive in structural engineering, because designers assume that ductile fracture

L.-J. Jia and H. Ge, *Ultra-low-Cycle Fatigue Failure of Metal Structures under Strong Earthquakes*, Springer Tracts in Civil Engineering,
https://doi.org/10.1007/978-981-13-2661-5_4

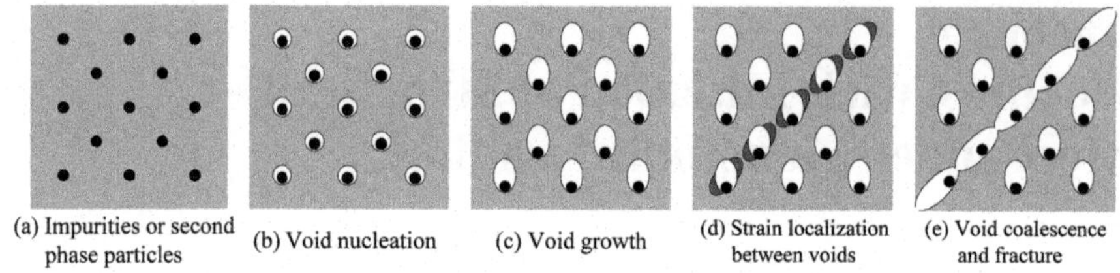

Fig. 4.1 Process of ductile fracture. Adapted from Anderson (2005)

can be prevented by the design rule such that joint strength be larger than the plastic strength of connected members, whose ultimate state is governed by buckling. However, the ductile fracture following buckling is still anticipated.

In practice, there are some limitations for the application of elaborate ductile fracture models. In structural engineering, designers commonly can only obtain monotonic tension coupon test results of steels. It is often not practical to require structural engineers to carry out complicated and elaborate experiments to calibrate a large number of model parameters related with the elaborate fracture models.

Several approaches have been proposed for the prediction of ductile fracture of steels, and a number of fracture models have been proposed, where some elaborate models are superior in the simulation accuracy but require calibration of many model parameters through complicated processes, and some empirical models may have relatively poor accuracy but is convenient to apply with limited number of model parameters and straightforward process for the calibration of the parameters. Herein, the features of these models are reviewed, from which the most promising model for application in structural engineering will be selected under the consideration of both accuracy of prediction results, and convenience in calibration of the model parameters in this chapter.

4.1.2 Approaches to Predict Ductile Fracture

(a) *Fracture Models Based on Void Growth and Void Coalescence*
Ductile fracture of metals commonly includes the following critical stages (Anderson 2005) as illustrated in Fig. 4.1.

Stage 1: Void nucleation at an impurity or second-phase particle as illustrated in Fig. 4.2 by either interface decohesion or particle cracking.
Stage 2: Void growth depending on both the equivalent plastic strain and hydrostatic stress.
Stage 3: Void coalescence when the voids grow to a critical size.

Some ductile fracture models have been proposed based on void growth and void coalescence. Mathematical models for void growth have been first proposed,

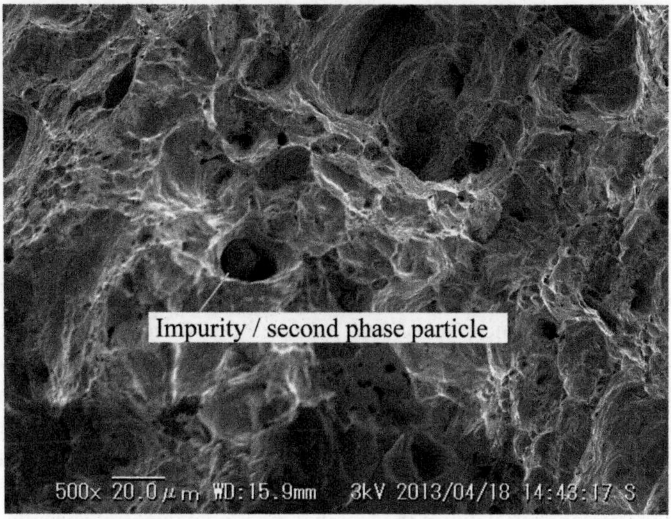

Fig. 4.2 Inclusion of impurity or second-phase particle in structural steel

among which the most widely referenced ones are the following two models, i.e., the McClintock model (Mcclintock 1968) and the Rice–Tracey model (Rice and Tracey 1969), where McClintock, Rice, and Tracey found the important role of hydrostatic stress on the growth of micro-voids.

McClintock proposed a criterion for ductile fracture by the growth of holes, and the model assumes two long cylindrical holes with elliptical cross section with axes parallel to the principle directions of the applied stress. The criterion of ductile fracture is that each hole touches a pair of its cell walls as the growth of the holes. A relative growth factor giving the increase in semi-axis of the hole relative to the corresponding hole spacing was proposed to define the crack initiation. The fracture model assumes a power law for the material and has two model parameters, where one is the stress hardening ratio of the material, and the other one is the equivalent strain at fracture. The model is convenient to apply in practice, while the assumption of the long cylindrical configuration of the voids is not realistic.

Rice and Tracey analyzed a more realistic model, i.e., a spherical void in a remote simple tension strain rate field and found that the fracture ductility (strain) decreased rapidly with the increase of stress triaxiality. They also found that the growth rate of the micro-void was an exponential function with the same coefficient in the exponential term for a non-hardening material subjected to tension for the case of a spherical void. The Rice–Tracey theory can describe the relationship between the growth rate of micro-voids and stress triaxiality straightforwardly with only one model parameter. However, the theory does not have a criterion for ductile fracture.

(b) *Fracture Models Based on Porous Plasticity*

Gurson (1975) proposed a porous plasticity constitutive model for ductile materials based on analysis results of a spherical void centered in a spherical cell, where the effect of hydrostatic stress on the yield function of the materials is considered. The plasticity model is quite different from common metal plasticity models, since the yield functions of the common ones are independent of hydrostatic stress. Rules for void nucleation, void growth, and void coalescence were also derived by postulating rigid-plastic materials. The condition of fracture was defined through a scalar named void volume fraction; i.e., crack forms when the parameter reaches a critical value. The void volume fraction was assumed due to void nucleation and void growth. Therefore, the Gurson model is a coupled material plasticity model and a ductile fracture model. Rudnicki and Rice (1975) and Yamamoto (1978) pointed out that there might be a possibility that ductile fracture can be induced by void growth within a localized narrow shear band, which can be resulted by initial imperfection of a material. From this point, the fracture strain sometimes might be greatly overestimated by the Gurson model due to the fact that the model does not consider the shear localization failure mode. Thus, Tvergaard (1981, 1982a, b) modified the Gurson model by adding three model parameters into the yield function of the model, and the rigid-plastic material model was also replaced by a power law material model with an isotropic hardening rule. The void growth rate close to the point of fracture initiation was found to accelerate in tension coupon tests (Tvergaard 1982a, b). Tvergaard and Needleman (1984) modified the evolution rule of the void volume fraction by introducing a failure point, after which the void growth rate accelerates. The modified model is commonly called Gurson–Tvergaard–Needleman model (GTN model). The GTN model can characterize void nucleation, growth, and coalescence, while the model has over ten parameters to determine for a single material. It may be difficult to be applied in structural engineering, since commonly only test results of smooth round coupons (or flat bars) under monotonic tension is available in practice. Another limitation of the GTN model is that the hardening rule is an isotropic hardening rule with a power law function, which commonly overestimates the stresses in reversal loading cycles, and also cannot simulate the Bauschinger effect which is a typical characteristic of structural steels. The isotropic hardening rule is known to be too rough to predict cyclic plasticity of structural steels, which is not proper to apply to simulation of ductile fracture of structural steels under cyclic large strain loading.

(c) *Fracture Models Based on Continuum Damage Mechanics*

The microscopic ductile fracture models are commonly at a scale level around 10^{-2}–10^{-1} mm, while a real metal structure is at a level of around 10^{2}–10^{3} mm. Continuum damage mechanics (CDM) provides another approach to simulate ductile fracture at the macroscopic level. The CDM is initiated from the concept of effective density of cracks or cavity over a plane at the macroscale level, and Kachanov (1958) firstly proposed a macroscopic damage index for the prediction of creep fracture. Chaboche (1984) and Lemaitre (1985) then formulated the constitutive equations of the CDM in the framework of thermodynamics, which provides scientific basis for the theory. Lemaitre proposed a CDM model for ductile fracture based on the concept of effective stress, which requires identification of three model parameters. Macroscopic crack initiates when the damage index, D, reaches a critical value. The model was also validated by comparison with the Rice–Tracey model. However, the calibration of the damage index, D, sometimes requires cyclic tests or measurements of the unloading Young's modulus at different strain levels using tension coupon tests, which is commonly inconvenient for the application to structural engineering.

4.1.3 Study on Ductile Fracture in Structural Engineering

Study on ductile fracture in structural engineering is quite limited. A simple fracture rule based on test results of various structural steels was first proposed by Kuwamura and Yamamoto (1997). Qian et al. (2005) simulated ductile fracture of circular hollow section joints under monotonic loading using the Gurson model. Kanvinde and Deierlein (2006) proposed a semiempirical stress modified critical strain (SMCS) model, which defines a scalar as the damage index, and it was also compared with the prediction results for monotonic test results of seven different structural steels using void growth model (VGM) based on the Rice–Tracey void growth theory. Smoothly notched tension coupon tests were required for the identification of the corresponding model parameter. Myers et al. (2010) studied the size dependency and empirical identification method of the model parameter of the SMCS model. The model was also applied to prediction of large-scale connections and structural members (Chi et al. 2006; Kanvinde and Deierlein 2007).

4.1.4 Approach Employed in This Chapter

The available approaches for predicting ductile fracture have been briefly reviewed. Considering that the aim of this chapter is to predict ductile fracture of structural members at macroscopic level using only tension coupon test results, a promising fracture model has to meet the following two aspects; i.e., the model can describe the microscopic mechanism of ductile fracture, and the model parameters can be simply obtained from a tension coupon test. It has been pointed out that the elaborate GTN model cannot well predict the cyclic plasticity of structural steels due to its isotropic hardening rule. Meanwhile, compared with other similar microscopic models, the void growth model proposed by Rice and Tracey is more widely accepted and is often applied to describe the void growth of ductile metals, which is commonly the critical step during ductile fracture. Therefore, a simple one-parameter semiempirical fracture model similar to the concept of Kachanov (1958) is proposed in this chapter, where the microscopic mechanism of ductile fracture is considered using the Rice–Tracey model.

Crack initiation is determined by definition of a crack. Despite the fact that a micro-crack is defined as a crack of size 0.01 mm, a crack in damage mechanics is commonly accepted as one of 1 mm (Chaboche 1988). Since a crack of 1 mm can be observed by the human eye as well as a digital camera, the damage mechanics definition is employed in this study.

In this chapter, a monotonic fracture model based on the concept of void growth is first proposed by applying the Rice–Tracey model and the Miner's rule (Miner 1945) in incremental form (Jia and Kuwamura 2014). A simple scalar damage index, *D*, is defined, and ductile fracture is postulated to occur when *D* reaches one. The fracture model is convenient to apply, since only one model parameter is required, which can be obtained from a tension coupon test straightforwardly. Then, the monotonic fracture model is validated through comparison between experimental and numerical simulation results. For the numerical simulation of ductile fracture under monotonic loading, the true stress–true strain data after necking initiation of the corresponding coupon tests is first modified using the modified weighted average method (Jia and Kuwamura 2014) proposed in Chap. 2, and then, numerical simulations of smooth and notched round coupons of three types of structural steels are carried out using the modified true stress–true strain data. The fracture model is found to predict well the load versus deformation curves and the fracture locations of the experiments.

4.2 Ductile Fracture Model under Monotonic Loading

4.2.1 Rice–Tracey Model

Rice and Tracey formulated the relationship between the radius of a void and stress triaxiality based on the analysis of a spherical void in a remote simple tension strain rate field as illustrated in Fig. 4.3. They found that the void growth rate could be approximated by the following formula for Mises materials.

$$\frac{\mathrm{d}R}{R} = 0.283 \cdot e^{\frac{3}{2}\frac{\sigma_h}{\sigma_{\mathrm{eq}}}} \mathrm{d}\varepsilon_{\mathrm{eq}} = 0.283 \cdot e^{\frac{3}{2}T} \mathrm{d}\varepsilon_{\mathrm{eq}} \tag{4.1}$$

where R is the radius of the void, σ_h and σ_{eq} are hydrostatic stress and equivalent stress, respectively, T is stress triaxiality, and $\varepsilon_{\mathrm{eq}}$ is equivalent strain.

4.2.2 Miner's Rule

Miner proposed a linear rule to evaluate cumulative damage under various loading histories of stress ratios, where the stress ratio is defined as the minimum stress to the maximum stress in one cycle of loading in a fatigue test. He assumed that there are M different loading cycles of various stress ratios, S_i, each with n_i cycles. If the

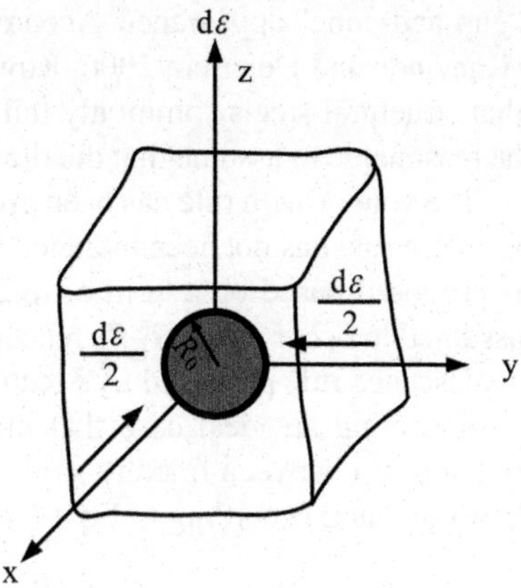

Fig. 4.3 Illustration of Rice–Tracey model. Adapted from Rice and Tracey (1969)

material under loading stress ratio of S_i fails under N_i cycles, then failure of the material will occur if the following criterion is satisfied.

$$\sum_{i=1}^{M} \frac{n_i}{N_i} = 1 \tag{4.2}$$

4.2.3 *Proposed Fracture Model for Monotonic Tension*

The Rice–Tracey model can be applied to describe the void growth, while no criterion for void coalescence is given. Integrating Eq. (4.1), you can obtain

$$\mathbf{ln}\frac{R}{R_0} = 0.283 \int_0^{\varepsilon_{eq}} \mathbf{e}^{\frac{3}{2}T} \mathrm{d}\varepsilon_{\mathrm{eq}} \tag{4.3}$$

where R_0 and R are the initial and current radii of the void, respectively. Assuming an ideal case that the stress triaxiality, T, is constant during the void growth, the equivalent strain can be given according to Eq. (4.3) as

$$\varepsilon_{eq} = \mathbf{ln}\frac{R}{R_0}/(0.283\mathbf{e}^{\frac{3}{2}T}) \tag{4.4}$$

Depending on the internal crystal structures of metals, the critical steps controlling the ductile fracture are different. For metals where the impurities and second-phase particles are well bonded to the matrix, void nucleation is often the critical step, and ductile fracture occurs soon after the voids form. Whereas for the metals where voids can nucleate easily, ductile fracture is commonly controlled by both void nucleation and void coalescence, and the fracture surfaces of the metals are characterized by the "cup-and-cone" appearance. According to the test results of several structural steels (Kanvinde and Deierlein 2006; Kuwamura and Yamamoto 1997), it has been found that structural steels commonly fail due to the void coalescence. Therefore, it may be reasonable to assume that ductile fracture occurs when void coalescence initiates.

The void growth rule has been given by the Rice–Tracey model, while a void coalescence rule has not been included in the model. Herein, a rule for void coalescence is proposed based on a term of relative growth factor, and the void coalescence is assumed to occur when R/R_0 reaches a critical value. The rule is similar to the void coalescence rule proposed by Mcclintock (1968).

Assuming an ideal case that the stress triaxiality is constant till fracture, the relationship between fracture strain (equivalent strain) and the stress triaxiality can be formulated according to Eq. (4.4)

$$\varepsilon_f = \mathbf{ln}\frac{R_f}{R_0}/(0.283\mathbf{e}^{\frac{3}{2}T}) = \chi \cdot \mathbf{e}^{-\frac{3}{2}T} \tag{4.5}$$

where ε_f and R_f are equivalent fracture strain and the radius of void when fracture occurs, respectively, and χ is a model parameter defining the critical value for the relative growth factor.

In order to extend the above equation to the case of general loading associates with non-constant T, a damage index, D, based on the Miner's rule in incremental form is proposed, where T is postulated to be constant during a single incremental step. The damage due to incremental strain $\mathrm{d}\varepsilon_{\mathrm{eq}}$ can be defined according to Eqs. (4.2) and (4.5)

$$\mathrm{d}D = \frac{\mathrm{d}\varepsilon_{\mathrm{eq}}}{\varepsilon_f(T)} = \frac{\mathrm{d}\varepsilon_{\mathrm{eq}}}{\chi \cdot \mathbf{e}^{-\frac{3}{2}T}} \tag{4.6}$$

where D is a damage index and a material is assumed to fracture when D reaches 1. Assuming that damage is only due to plastic deformation, Eq. (4.6) can be written as

$$\mathrm{d}D \approx \frac{\mathrm{d}\varepsilon_{\mathrm{eq}}^{p}}{\chi \cdot \mathrm{e}^{-\frac{3}{2}T}} \tag{4.7}$$

4.2.4 Calibration of Model Parameter

For the proposed fracture model, there is only one model parameter, χ, which can be calibrated from a monotonic tension coupon test. Numerical analyses of a coupon under monotonic tension test are necessary to obtain χ according to the following process:

Step 1: Obtain the true stress and true strain data of the material using the modified weighted average method introduced in Chap. 2.

Step 2: Give an initial value of χ, e.g., $\chi = 2.0$, and carry out a numerical simulation of the coupon test using the material data obtained in Step 1.

Step 3: Compare the fracture point of the experimental load–deformation curve with that of the simulation result.

If the fracture point of the numerical result in Step 3 compares well with that of the test result, then the χ is the optimal value. If the comparison result is not good, go back to Step 2 and give a new χ based on the comparison result, and repeat the loop until the comparison result is acceptable.

4.3 Experiment

To study the effect of stress triaxiality on ductile fracture of structural steels, three series of round coupons made of JIS SS400, HT800 (Kuwamura and Yamamoto 1997) and SM490 (Arita and Iyama 2009) with the configurations as shown in Figs. 4.4 and 4.5, respectively, were monotonically pulled to fracture. The mechanical properties and chemical compositions are given in Table 4.1. The tests were carried out in room temperature under displacement control at a quasi-static loading rate. The test setups of the three steels are similar to that of the hourglass-type specimens in Chap. 3. The extension within a gage length of 30 mm was also measured by an extensometer for the tests done by Arita and Iyama, and the minimum radii of the coupons were measured using a digital micrometer for both of the two sets of tests. Typical "cup-and-cone" type fracture was found in all the coupons. Crack initiates at the central regions of the minimum cross sections for Type 4, Type 5, and Type 6, while cracks start from the surfaces of the notch roots for the other three coupons with sharper notches. The locations of crack initiation for a smooth round bar and a sharply-notched one are illustrated in Fig. 4.6.

4.4 Numerical Simulation

4.4.1 FE Modeling

The ductile fracture of the smooth and notched coupons is simulated by element deletion in ABAQUS/Explicit (ABAQUS 2010), where the element is removed when the damage index D reaches 1. Axisymmetric models using CAX8 elements are applied to carry out the quasi-static simulations of the tests. The mesh for Type 6 is illustrated in Fig. 4.7. Displacement is applied to the left edge, and the right edge is restrained along the axis direction.

Table 4.1 Mechanical properties and chemical composition of steels

Steel	Mechanical properties						Chemical composition (Weight %)				
	Yield stress (MPa)	Tensile strength (MPa)	σ_{neck} (MPa)	ε_{neck}	w	χ	C	Si	Mn	P	S
SS400	270	460	565	0.21	0.37	1.8	0.17	0.27	0.72	0.018	0.009
HT800	763	826	896	0.07	0.23	1.9	0.20	0.25	1.41	0.11	0.006
SM490	401	546	640	0.20	0.32	2.4	0.13	0.27	1.28	0.02	0.006

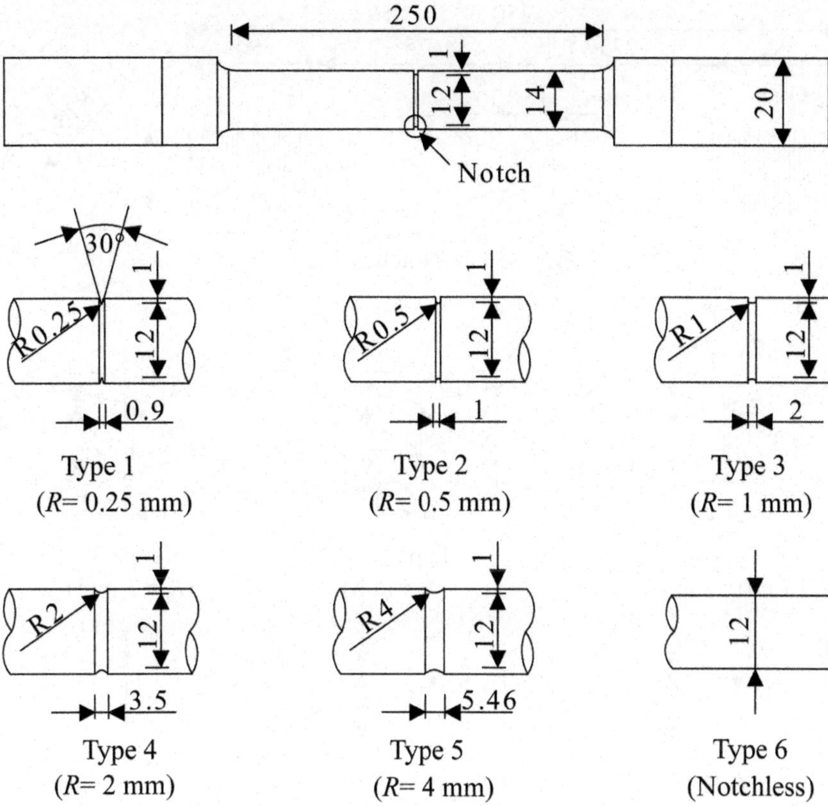

Fig. 4.4 Configuration of round and notched coupons (SS400 and HT800)

4.4.2 Plasticity Model and Calibration of Model Parameters

Herein, the nonlinear isotropic hardening model was employed in the simulation, since the model is accurate enough to describe metal plasticity under monotonic loading. True stress–true strain data till fracture is required to be inputted in tabular form into ABAQUS, which can be obtained by the MWA method proposed in Chap. 2 using the test results of the smooth coupons (Type 6). The parameters of the MWA method are as shown in Table 4.1.

Ductility of a material is not just decided by the value of the model parameter, χ, but also depended on the stress–strain behavior. For a tension coupon test, ductility of a material is commonly evaluated by the elongation within a gage length, and the elongation can be divided into two parts, i.e., the elongation before necking occurs and that after necking initiates. The first part is decided mainly by the true strain corresponding to the initiation of necking, Cracks occur at a quite early stage for, and the second part is mainly depended on both the fracture parameter, χ, and the

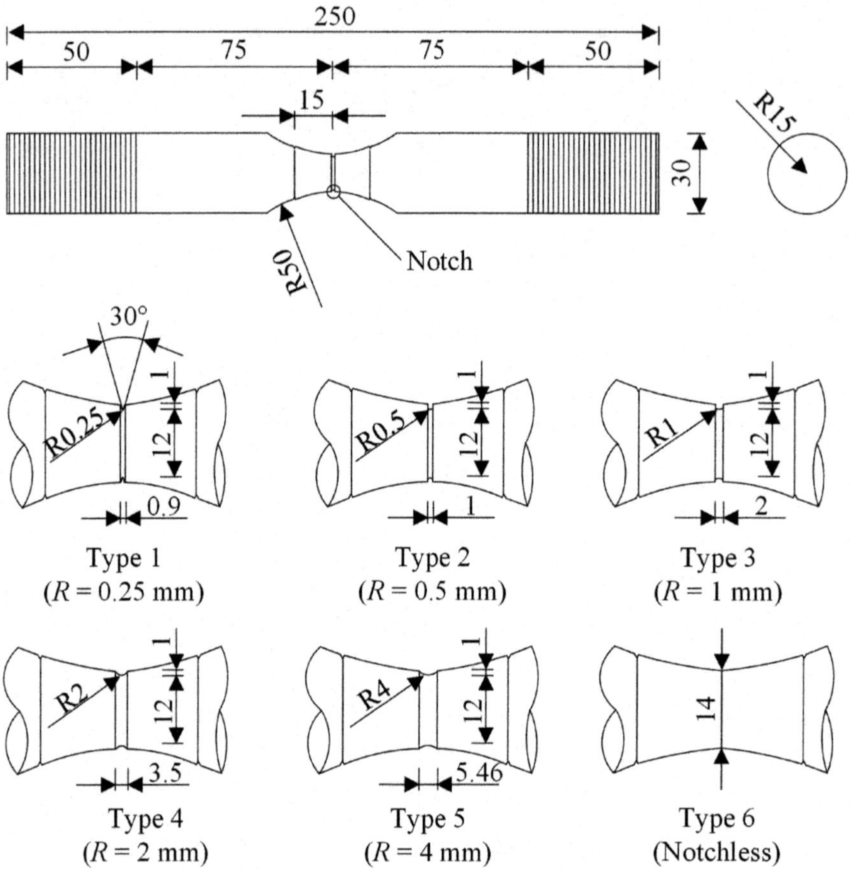

Fig. 4.5 Configuration of round and notched coupons (SM490)

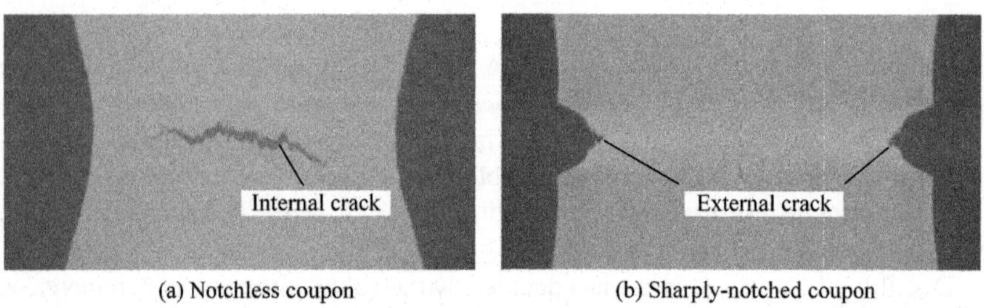

Fig. 4.6 Locations of crack initiation for coupons with different shapes (provide by Funabashi Shingo)

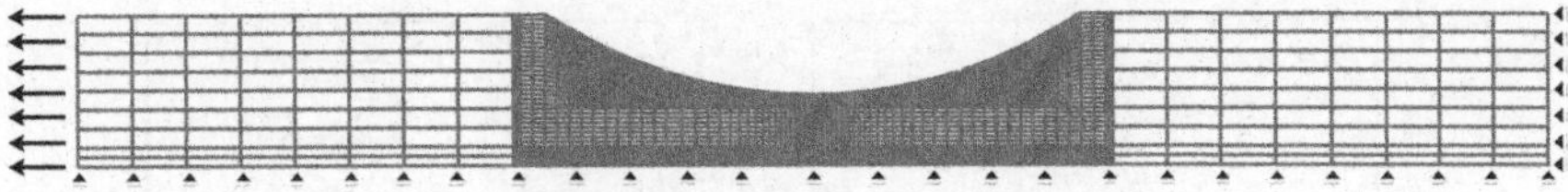

Fig. 4.7 FE model of Type 6 of the tests carried out by Arita and Iyama

hardening modulus of the material. The χ is more concerned with the localized deformation within the necked region, and a large value of χ indicates a significant reduction in the cross-sectional area of the minimum cross section of a coupon. Therefore, it is not strange that though the elongation of the high-strength steel HT800 is smaller than that of the mild steel SS400 according to the coupon test results, while the fracture parameter, χ, of the high-strength steel HT800 is larger than that of the mild steel SS400. It can be found from the table that the true strain at the initiation of necking of SS400 is much larger than that of HT800, which indicates that the elongation before the initiation of necking of SS400 is much larger. It is also interesting to find that the hardening modulus after necking occurs, which can be obtained as $w \cdot \sigma_{\text{neck}}$, is close to a constant of 206 MPa for the tested steels. Herein, the true stress and true strain are calculated based on the measured instantaneous diameter of the minimum cross section using a digital laser micrometer, since it is known that the localized deformation data is more accurate. The true stress and true strain calibrated by the elongation of an extensometer may give different conclusions.

4.5 Comparison of Experiment and Simulation

Numerical simulation results of the notched round coupons of the three steels are shown in Figs. 4.8 and 4.9, where the load–displacement curves and displacements at crack initiation agree well with those of the test results. Cracks occur at a quite early stage for the sharply-notched coupons (Types 1 and 2); e.g., crack initiates near the peak of the load–deformation curve for Type 1. The predicted locations where cracks initiate of SM490 are shown in Fig. 4.10, where cracks initiate from the surfaces of the coupons for sharply-notched coupons (Types 1–3), while cracks occur at the centers for bluntly-notched coupons. The numerical results indicate that the shaper the notch is, the smaller the yield region will be. The distribution of plastic strain at fracture is quite uniform for bluntly-notched specimens, e.g., Type 6, while the plastic strain is concentrated nearby the notch root for the sharply-notched specimens, e.g., Type 1. The predicted locations of crack initiation are also in good accordance with the test results.

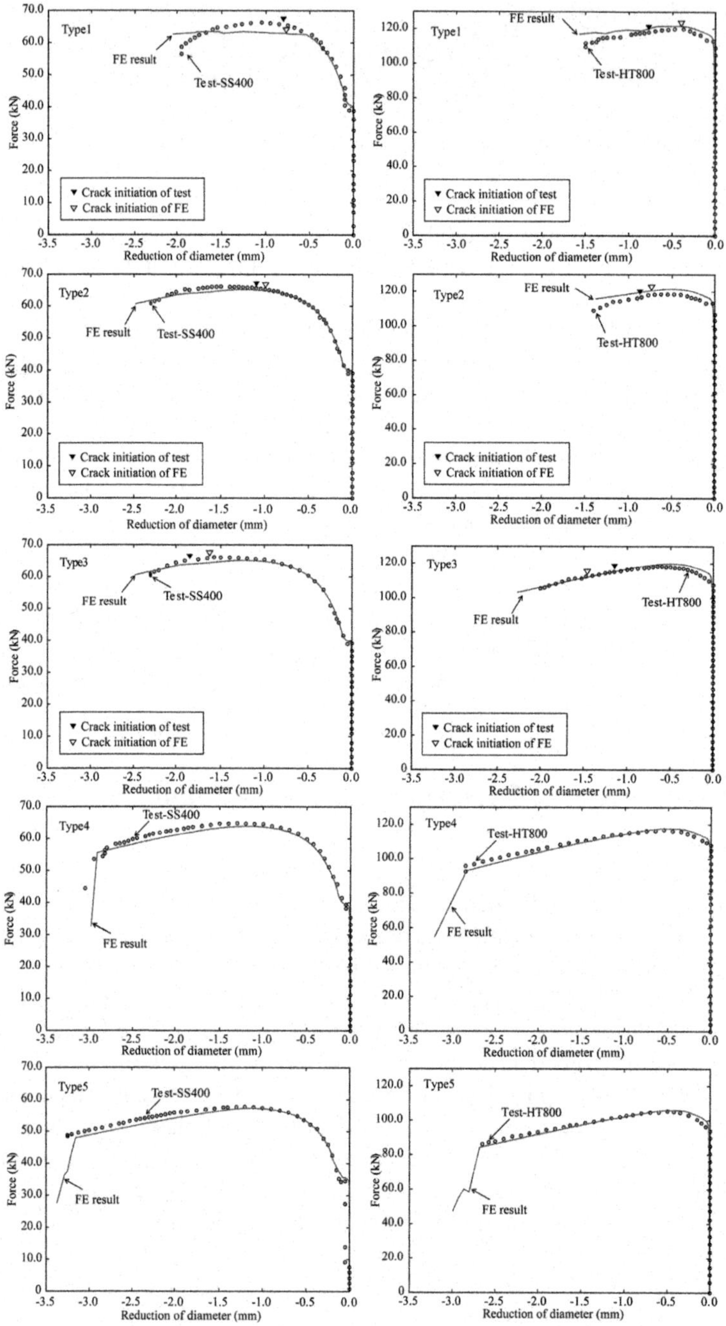

Fig. 4.8 Comparison between experimental and numerical results of SS400 and HT800

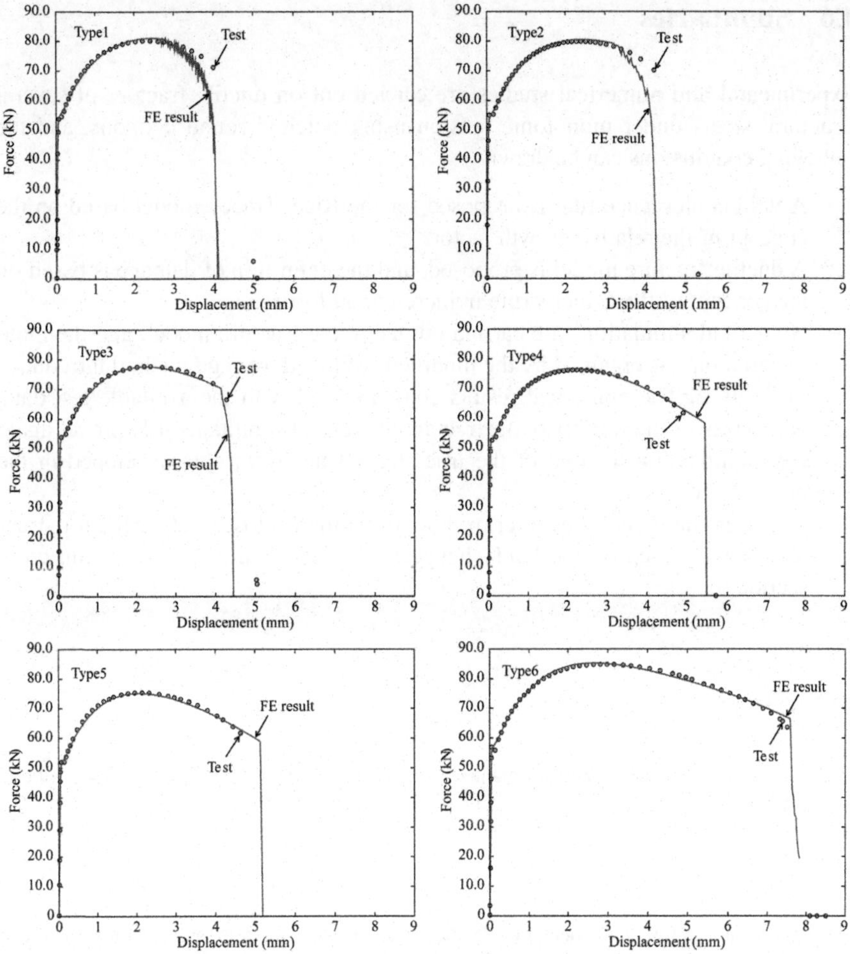

Fig. 4.9 Comparison between experimental and numerical results of SM490

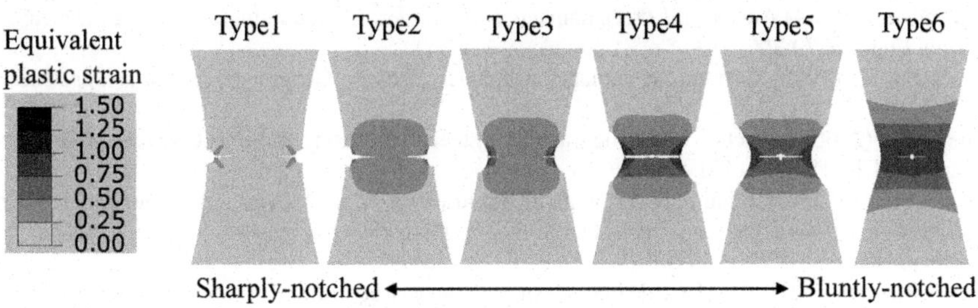

Fig. 4.10 Locations of crack initiation predicted by FE results (SM490)

4.6 Summaries

Experimental and numerical studies are carried out on ductile fracture of several structural steels under monotonic tension using notched round coupons, and the following conclusions can be drawn

(1) A void coalescence rule is proposed for the Rice–Tracey model based on the concept of the relative growth factor.
(2) A ductile fracture model is proposed, and the definition of damage is based on the concept of the Miner's rule in incremental form.
(3) Numerical simulations are carried out using the fracture model, and the input material data is obtained by the modified weighted average method introduced in Chap. 2. The numerical results compare well with the available test data, which proves the validity of the fracture model and simultaneously proves again the modification method of the true stress–true strain data developed in the preceding chapter.
(4) The stress hardening rate after necking initiation is close to 206 MPa for structural steels based on the load–change in dimeter data of the minimum cross section.

References

ABAQUS (2010) ABAQUS standard manual (version 6.10). Karlsson & Sorensen Inc., Hibbitt, Pawtucket

AIJ (1995) Fracture in steel structures during a severe earthquake. Architectural Institute of Japan, Tokyo

Anderson TL (2005) Fracture mechanics: fundamentals and applications. Taylor and Francis, London

Arita M, Iyama J (2009) Fundamental verification of crack extension rule for estimating fracture of steel. In: Proceedings of the summaries of technical papers of meeting Architectural Institute of Japan B-1, structures I, Loads, reliability stress analyses foundation structures shell structures, space frames and membrane structures, pp 273–274

Chaboche JL (1984) Anisotropic creep damage in the framework of continuum damage mechanics. Nucl Eng Des 79:309–319

Chaboche JL (1988) Continuum damage mechanics: Part I—General concepts. J Appl Mech 55:59–64

Chi W et al (2006) Prediction of ductile fracture in steel connections using SMCS criterion. J Struct Eng (ASCE) 132:171–181

Gurson AL (1975) Continuum theory of ductile rupture by void nucleation and growth. Part I. Yield criteria and flow rules for porous ductile media. Division of Engineering, Brown University, Providence

Jia L-J, Kuwamura H (2014) Ductile fracture simulation of structural steels under monotonic tension. J Struct Eng (ASCE) 5(140):04013115

Kachanov LM (1958) Time of the rupture process under creep conditions. Izvestiya Akademii Nauk SSSR Otdelenie Tekniches 8:26–31

Kanvinde A, Deierlein G (2006) The void growth model and the stress modified critical strain model to predict ductile fracture in structural steels. J Struct Eng (ASCE) 132:1907–1918

Kanvinde A, Deierlein G (2007) Finite-element simulation of ductile fracture in reduced section pull-plates using micromechanics-based fracture models. J Struct Eng (ASCE) 133:656–664
Kuwamura H (1998) Fracture of steel during an earthquake—state-of-the-art in Japan. Eng Struct 20:310–322
Kuwamura H (2003) Classification of material and welding in fracture consideration of seismic steel frames. Eng Struct 25:547–563
Kuwamura H, Akiyama H (1994) Brittle fracture under repeated high stresses. J Constr Steel Res 29:5–19
Kuwamura H, Yamamoto K (1997) Ductile crack as trigger of brittle fracture in steel. J Struct Eng (ASCE) 123:729–735
Kuwamura H et al (2003) Effects of material toughness and plate thickness on brittle fracture of steel members. J Struct Eng (ASCE) 129:1475–1483
Kuwamura Lab (1995) Field survey report on structural damage during the 1995 Hyogoken-Nanbu Earthquake. Kuwamura Lab, School of Engineering, The University of Tokyo, Tokyo
Lemaitre J (1985) A continuum damage mechanics model for ductile fracture. J Eng Mater Technol 107:83–89
Mahin SA (1998) Lessons from damage to steel buildings during the Northridge earthquake. Eng Struct 20:261–270
McClintock FA (1968) A criterion for ductile fracture by the growth of holes. J Appl Mech 35:363–371
Miner MA (1945) Cumulative damage in fatigue. J Appl Mech 12:A159–A164
Myers A et al (2010) Calibration of the SMCS criterion for ductile fracture in steels: specimen size dependence and parameter assessment. J Eng Mech (ASCE) 136:1401–1410
Qian X et al (2005) Simulation of ductile fracture of circular hollow section joints using the Gurson model. J Struct Eng (ASCE) 131:768–780
Rice JR, Tracey DM (1969) On the ductile enlargement of voids in triaxial stress fields. J Mech Phys Solids 17:201–217
Rudnicki JW, Rice JR (1975) Conditions for the localization of deformation in pressure-sensitive dilatant materials. J Mech Phys Solids 23:371–394
The Building Center of Japan (2003) Guidelines for prevention of brittle fracture at the beam ends of welded beam-to-column connections in steel frames. The Building Center of Japan, Tokyo
Tvergaard V (1981) Influence of voids on shear band instabilities under plane strain conditions. Int J Fract 17:389–407
Tvergaard V (1982a) Material failure by void coalescence in localized shear bands. Int J Solids Struct 18:659–672
Tvergaard V (1982b) On localization in ductile materials containing spherical voids. Int J Fract 18:237–252
Tvergaard V, Needleman A (1984) Analysis of the cup-cone fracture in a round tensile bar. Acta Metall 32:157–169
Yamamoto H (1978) Conditions for shear localization in the ductile fracture of void-containing materials. Int J Fract 14:347–365

Chapter 5
Ductile Crack Propagation under Monotonic Loading

Abstract For specimens with relatively uniform stress and strain distribution, e.g., smooth tension coupons, a crack propagates almost simultaneously as the crack initiates. For most of the cases, a ductile crack propagation rule is required to accurately evaluate the whole cracking process. In this chapter, a crack propagation rule based on an energy balance approach is added to the crack initiation rule in the previous chapter, where deterioration of material is also considered. A method to calibrate the crack propagation correlated parameter using tension coupon test results is proposed. A series of single-edge V-notched and U-notched tension coupons made of high-strength steel are manufactured and experimentally pulled to rupture under quasi-static loading. Numerical cracking simulation is also conducted and compared with the experimental results. Comparison results indicate that the ductile fracture model can well predict both crack initiation and crack propagation of structural steel under high triaxialities.

5.1 Introduction

Subsequent progressive brittle fracture after ductile cracking (Kuwamura and Yamamoto 1997) of welded connections in steel moment resisting frames (SMRF) has been observed during the 1994 Northridge and the 1995 Kobe earthquakes (AIJ 1995; Bruneau et al. 1996; Mahin 1998; O'sullivan et al. 1998). A number of studies have been conducted to clarify the failure mechanisms of the aforementioned failure modes (e.g., Huang et al. 2008; Kanvinde and Deierlein 2006; Khandelwal and El-Tawil 2014; Kuwamura and Yamamoto 1997; Mackenzie et al. 1977; Panontin and Sheppard 1995; Rousselier 1987) and to improve seismic performance of the welded connections (e.g., Gilton and Uang 2002; Kim et al. 2002; Sumner and Murray 2002), after the two strong earthquakes.

It is of great importance to evaluate the ductile cracking phenomenon preceding the brittle fracture. Besides, ductile fracture of steel members under cyclic large plastic strain loading was also reported during the 1995 Kobe earthquake, i.e., ductile cracking of steel bracings in a garage as shown in Figs. 1.2. Up to date, it is still

L.-J. Jia and H. Ge, *Ultra-low-Cycle Fatigue Failure of Metal Structures under Strong Earthquakes*, Springer Tracts in Civil Engineering,
https://doi.org/10.1007/978-981-13-2661-5_5

difficult to accurately evaluate the whole ductile cracking process of structural steel due to the fact that ductile fracture is an extremely nonlinear phenomenon, where extremely large plasticity and geometrical deformation are involved. Additionally, the analysis objects in structural engineering are commonly uncracked bodies. This is different from that of conventional linear fracture mechanics problems, where pre-cracked bodies are treated.

A number of ductile crack initiation models termed void growth models (VGM) (Jia and Kuwamura 2014; Jia and Kuwamura 2015; Kanvinde and Deierlein 2006; Panontin and Sheppard 1995; Rousselier 1987; Zhou et al. 2013) were proposed to evaluate crack initiation of metals, and this category of ductile fracture models are based on the concept of void growth (Mcclintock 1968; Rice and Tracey 1969). Ductile crack is postulated to initiate when voids reach a critical size and coalescence with each other (Anderson 2005). VGM can generally evaluate ductile crack initiation of structural steel and structural members with good accuracy (Jia et al. 2014; Jia and Kuwamura 2014; Kanvinde and Deierlein 2007; Kiran and Khandelwal 2014a, b; Roufegarinejad and Tremblay 2012), since structural steels are commonly ductile, where void growth and void coalescence are critical stages dominating ductile crack initiation.

To date, research on ductile crack propagation of metals without pre-cracks is still limited. However, the issue of ductile crack propagation is important for evaluation of strength, ductility, and energy dissipation capacity of metal structures, since load-carrying capacity sometimes decreases only when a substantial large crack is developed. For a VGM, ductile crack propagation is commonly postulated to occur simultaneously as the crack initiation, or the ductile crack initiation index reaches a threshold value over a critical length (Kanvinde and Deierlein 2006). Ductile crack initiation and propagation are commonly simulated by element deletion for initially uncracked solids, and an element is removed when the crack initiation index reaches unit. Acceptable accuracy can be obtained for these approaches when there is no large gradient for the crack initiation index, i.e., no large strain gradient or stress triaxiality gradient. For these cases, ductile crack propagates quickly after crack initiation. This can be observed during the final rupture of a smooth tension coupon, where a crack propagates immediately after crack initiation, even though the cracked surface is 100% ductile. Besides, these approaches can also simulate crack propagation of metals with low toughness, where brittle fracture will occur soon after ductile crack initiation. However, they cannot simulate well for the cases where large gradient for the crack initiation index is involved, such as bending, especially for high-strength steel with high toughness in structural engineering. A ductile fracture model with only a crack initiation rule generally overestimates the crack propagation rate, especially for the cases where non-uniform strain distribution prevails (e.g., Jia et al. 2016).

In structural engineering, mechanical properties of structural steels can significantly vary from each other, even for the same steel grade. Structural engineers commonly can only obtain tension coupon test results, and it is of significant practical value if one can obtain all the fracture parameters necessarily required by a ductile fracture model from common tension coupon test results. In the previous study of the authors, a ductile fracture model with both crack initiation and crack propagation rules was employed to simulate ductile cracking of structural steel under cyclic combined shear and normal stress loading, where the crack propagation rule is defined in terms of ductile fracture energy and the fracture energy is obtained from Charpy test results within the ductile–brittle fracture transition region. This chapter aims to propose and validate a straightforward method to obtain the ductile fracture energy only using tension coupon tests of high-strength steel under high stress triaxiality. Shear fracture-related cases (Barsoum and Faleskog 2007; Kiran and Khandelwal 2014a, b; Nahshon and Hutchinson 2008; Tvergaard 2009; Tvergaard and Nielsen 2010) with low and medium stress triaxialities are out of the scope of this chapter. A series of single-edge V-notched tension (SEVNT) and single-edge U-notched tension (SEUNT) specimens were manufactured and experimentally pulled to fracture under quasi-static loading. Numerical simulations were also carried out, where the parameters of crack initiation and crack propagation rules were both calibrated using common tension coupon tests. Comparison between the experimental and numerical results indicates that the newly proposed method to obtain the ductile fracture energy of the ductile crack propagation rule can well simulate both the crack initiation and propagation of the specimens without existing cracks.

5.2 Ductile Fracture Model

5.2.1 Crack Initiation Rule

A ductile crack initiation rule was proposed to simulate ductile fracture under cyclic large plastic strain loading (Jia et al. 2014), where a damage index D_{ini} for ductile crack initiation is defined in a small increment during numerical integration,

$$\mathrm{d}D_{\text{ini}} = \frac{\mathrm{d}\varepsilon_{\text{eq}}^{p}}{\chi_{\text{cr}} \cdot e^{-\frac{3}{2}T}} \tag{5.1}$$

where χ_{cr} = material parameter correlated with the crack initiation, $\mathrm{d}\varepsilon_{\text{eq}}^{p}$ = incremental equivalent plastic strain, T = stress triaxiality defined by the ratio of hydrostatic pressure to the equivalent Mises stress for metals. Postulating that damage accumulation under different stress triaxilities follows a linear rule, i.e., the Miner's rule (Miner 1945), ductile fracture initiation of a material is postulated to occur when D_{ini} reaches unit. The term "ductile fracture initiation" herein denotes a micro-crack at a scale of 0.01–0.1 mm. In previous studies, a fracture model with only the ductile crack initiation rule in Eq. (5.1) was employed to predict ductile cracking of structural steel (Jia et al. 2016) and steel members under cyclic loadings (Jia et al. 2014), where no crack propagation rule was employed, and cracks were assumed to propagate immediately after crack initiation when the damage index D_{ini} reached unit. Crack propagation was simulated by element deletion when the damage index D_{ini} at one of the integration points of the solid element reached unit. As stated in the introduction, this assumption may perform well when stress and strain distributions are uniform across the cross section. In the two aforementioned studies (Jia et al. 2014, 2016), crack propagation processes cannot be well evaluated using the ductile fracture model with only a crack initiation rule due to the fact that the stress and strain distributions are non-uniform, where flexural deformation is dominant, and the predicted moment of rupture using the one without a crack propagation rule is commonly earlier than those of the experiments.

5.2.2 *Crack Propagation Rule*

A ductile crack propagation rule based on an energy balance approach (Hillerborg et al. 1976) is employed in this study, and it is postulated that a certain amount of energy G_c is absorbed to generate a unit area of crack surface. During crack propagation, a certain amount of stored energy is released, and a crack propagates when the released energy is greater than the absorbed energy to open the crack. Combined with the VGM, this approach can simulate both crack initiation and crack propagation of uncracked bodies with relatively large mesh, while conventional fracture mechanics methods are only applicable to crack propagation of pre-cracked objects with fairly fine mesh at the crack tips (Qian and Yang 2012). A crack propagation index, D_{prop}, can be defined,

$$D_{\text{prop}} = \frac{G}{G_c} \tag{5.2}$$

where G is current absorbed energy of a unit area since the occurrence of fracture initiation as illustrated in Fig. 5.1, and G_c is the threshold value for absorbed energy of a unit area of crack surface. G_c can be calculated from the following equation using the correlation between the effective stress and crack width, w_1,

$$G_c = \int_0^{w_1} \sigma_e \text{d}w \tag{5.3}$$

where w_1 is crack width when D_{prop} reaches unit, and σ_e is the effective stress of a damaged material as illustrated in Fig. 5.1. Different from the original crack propagation rule proposed in the literature (Hillerborg et al. 1976), material deterioration is considered through the concept of effective stress in this study,

$$\sigma_e = (1 - D_{\text{prop}}) \cdot \sigma \tag{5.4}$$

where Eq. (5.4) postulates that the effective stress decreases linearly as D_{prop} increases. During an FE analysis, the crack width can be described by equivalent plastic strain using a concept of characteristic element length, l_c. The definition of

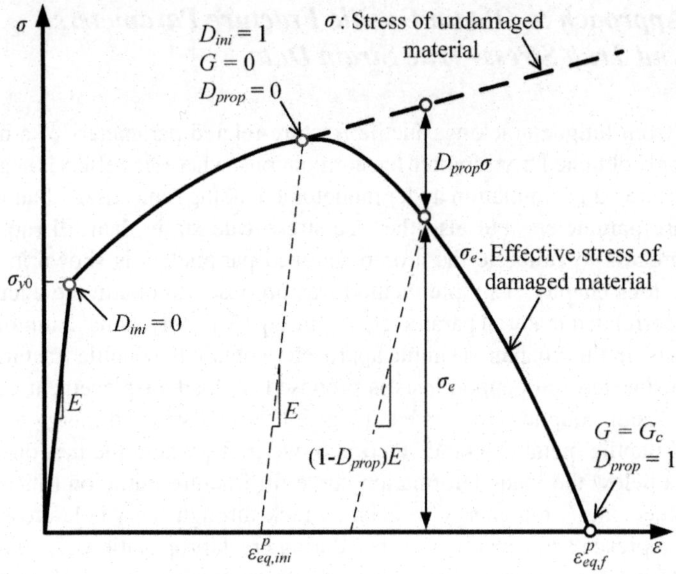

Fig. 5.1 Illustration of ductile fracture model

the characteristic element length depends on the element geometry and formulation. For beams and trusses, it is a characteristic length along the element axis. For plane elements, it is the square root of the area. For solid elements, it is the third root of the element volume. The damage index, D_{prop}, can also be given by

$$D_{\text{prop}} = \frac{l_c \cdot (\varepsilon^p_{\text{eq}} - \varepsilon^p_{\text{eq, ini}})}{w_1} \tag{5.5}$$

where $\varepsilon^p_{\text{eq}}$ is equivalent plastic strain, and uniaxially loaded tension coupon. Given$\varepsilon^p_{\text{eq, ini}}$ is equivalent plastic strain at fracture initiation when $\underline{D_{\text{ini}}}$ and D_{prop}, respectively, equal to 1.0 and 0 as shown in Fig. 5.1. This treatment can reduce mesh dependence of the FE method and thus make it possible to simulate ductile fracture problems with relatively large elements.

In addition, the loading and unloading tangent moduli of a damaged material, E_d, are also reduced in this study, which is considered in the original crack propagation rule.

$$E_d = (1 - D_{\text{prop}}) \cdot E \tag{5.6}$$

where E = initial tangent modulus of an undamaged material.

5.2.3 Approach to Obtain Ductile Fracture Parameters and True Stress–True Strain Data

Different from fatigue cracking, ductile fracture-related parameters of a metal can be commonly obtained from smooth tension coupons, where size effect is neglectable. For ductile cracking simulation under monotonic loading, one has to obtain not only the fracture parameters, but also the true stress–true strain data till rupture. The detailed process to calibrate the aforementioned parameters is shown in Fig. 5.2. In the previous chapter, a simple method was proposed to obtain the ductile crack initiation correlated material parameter, χ_{cr}, in Eq. (5.1) only using common tension coupon tests. In this chapter, a simple approach to obtain the ductile fracture energy, G_c, only using tension coupon tests is proposed. A load–displacement curve of a common tension coupon is illustrated in Fig. 5.3. Total dissipated energy of a tension coupon of ductile metal can be divided into two parts, where the first one is equal to the area below the load–deformation curve till fracture initiation (micro-crack), and the other one is the energy to open a crack through the whole cross section. The latter represents the energy absorbed during crack propagation, i.e., the fracture energy. During crack propagation stage, plastic straining only occurs at the crack tips while unloading occurs for the material at the other parts owing to decrease of the load-carrying capacity. If the tension coupon test is controlled by displacement loading under a slow speed and the sampling frequency is high enough, it is possible to capture the rupture process till the moment when the load decreases to 0. To capture the rupture process, the testing frame should also be stiff enough to ensure the rupture process quasi-static. This method can obtain an accurate value of G_c, since the decreasing load will not induce remarkable plasticity in other regions during the crack propagation stage. The shaded energy in Fig. 5.3 is the fracture energy of the minimum cross section. To calculate the threshold value for absorbed energy per unit area, G_c, one has to know the area of the deformed minimum cross section. Due to the irregular shape of the fracture surface, it is difficult to accurately measure the cross-sectional area. A method to estimate the cross-sectional area of the cracked surface is proposed in this study.

As is known, necking initiates when the peak load, P_u, is reached for a uniaxially loaded tension coupon. Given an initial cross-sectional area of A_0, and current cross-sectional area of A, the logarithmic strain (true strain) can be expressed as

$$\varepsilon = \ln\frac{A_0}{A} \tag{5.7}$$

Thus, the cross-sectional area corresponding to the peak load, A_u, can be obtained,

$$A_u = \frac{A_0}{e^{\varepsilon_u}} \tag{5.8}$$

where ε_u is the true strain corresponding to the peak load. To calculate the minimum cross-sectional area at the moment of rupture, two assumptions are made:

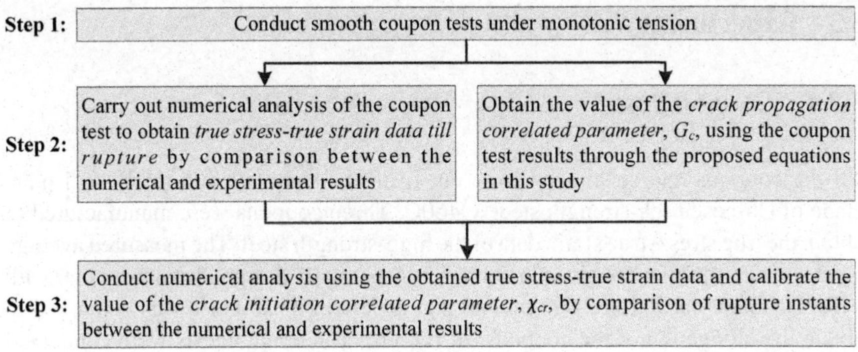

Fig. 5.2 Process to calibrate fracture model parameters and true stress–true strain data till rupture

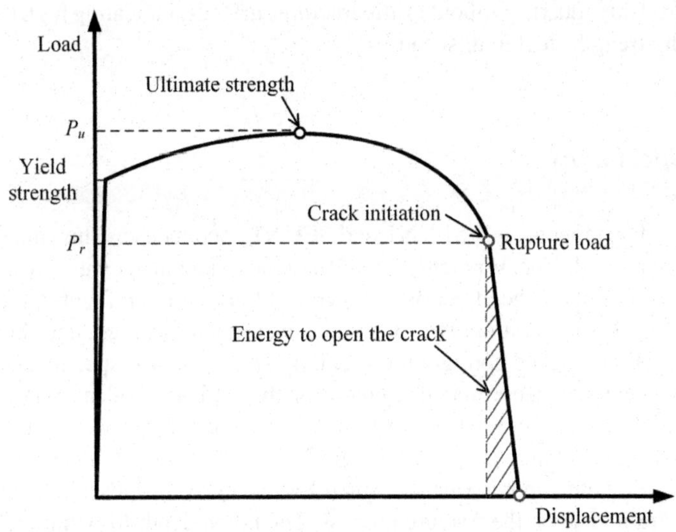

Fig. 5.3 Illustration of method to calculate fracture energy using tension coupon test

(1) No strain hardening occurs after the peak load is achieved.
(2) The cross-sectional area is proportional to the load after neck initiation.

Based on the two assumptions, the area of the minimum cross section corresponding to rupture, A_r, can be calculated

$$A_r = \frac{P_r}{P_u} A_u \tag{5.9}$$

where P_r is the load corresponding to rupture as illustrated in Fig. 5.3.

5.3 Experiment

5.3.1 Coupon Test

All the coupons and specimens were cut from the same 12-mm-thick steel plate made of Chinese high-strength steel, Q460C. Three coupons were manufactured to obtain the true stress–true strain data of the high-strength steel. The measured average mechanical properties are given in Table 5.1. A specially designed extensometer with a displacement capacity of 60 mm and a gage length of 200 mm was employed to obtain the elongation of the coupons till rupture. The coupon tests were conducted in room temperature under a quasi-static speed, where the speed is slow enough to capture the displacement data during rupture. It has been found that the fracture surfaces are 100% ductile without brittle fracture surfaces, indicating high toughness of the high-strength steel in this study.

5.3.2 Specimens

Two series of specimens, i.e., SEVNT and SEUNT specimens as shown in Fig. 5.4, were manufactured. The V-notched and U-notched specimens were, respectively, designed to investigate the effect of strain concentration on the failure process. For each V-notched configuration, two specimens were manufactured, and three specimens for each U-notched configuration. Totally six V-notched specimens and nine U-notched specimens were manufactured. For the limitation of mechanical workmanship, the radii at the notch root of the SEVNT specimens were set to be less than 0.1 mm. The notch radius of the SEUNT specimens was all designed as 2 mm. Three different notch depths were designed for each series to investigate the effect of plastic strain level on the fracture process. The notch depth-to-width ratios were, respectively, designed as 10, 30, and 50%. The widths of the uniform sections were all set as 40 mm. The numbering of the specimen is illustrated, e.g., "SEVNT-10-1", where SEVNT = single-edge V-notched specimens, 10 = the notch depth-to-width ratio of 10%, 1 = the first specimen of SEVNT-10.

Table 5.1 Measured average mechanical properties

Steel	Yield stress (MPa)	Tensile strength (MPa)	Elongation (%)	χ_{cr}	G_c (J/mm^2)
Q460C	480.6	560.7	15.8	2.4	0.22

Notes G_c = threshold fracture energy of material with a unit area since fracture initiation; χ_{cr} = material parameter related to the crack initiation rule

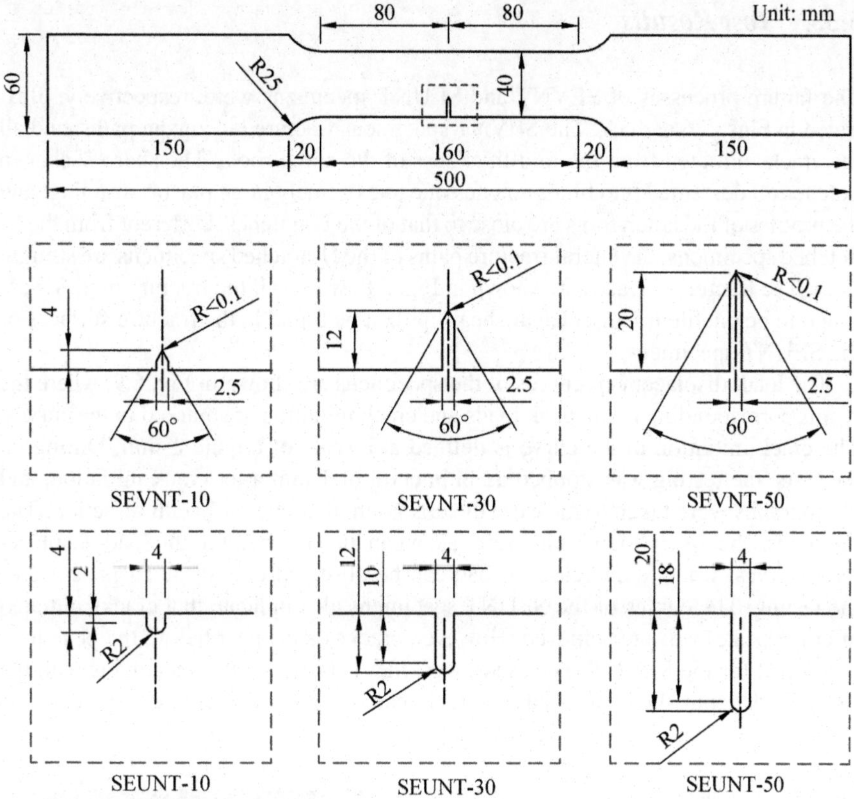

Fig. 5.4 Configurations of specimens

5.3.3 Test Setup

All the specimens were loaded using an MTS loading system with a force capacity of 500 kN and a stroke capacity of ±75 mm, where the reaction frame with four compact columns is stiff enough to ensure the rupture process of the tension coupon test quasi-static. The test setup is shown in Fig. 5.5, where four displacement transducers were employed to monitor the displacement. All the tests were controlled using the average net displacement data between the upper and lower displacement transducers. All the tests were conducted under room temperature, and all the tests were carried out at a quasi-static speed.

5.3.4 Test Results

The failure processes of SEVNT and SEUNT specimens were, respectively, illustrated in Figs. 5.6 and 5.7. The SEVNT specimens fractured in zigzag paths, and all the cracks initiated from the mid-thickness of the notch roots. The sharp V-shaped notches all deformed into blunt notches due to extremely large plastic straining, and the profiles of the notch roots are close to that of the U-notches. Different from the V-notched specimens, the lateral fracture paths of the U-notched specimens are straight lines. The fracture surfaces are shown in Fig. 5.8, where all the fracture modes were found to be ductile fracture. Clear shear lips can be found in the fracture surfaces of the SEUNT specimens.

The load–displacement curves of the specimens are shown in Fig. 5.9, where the points corresponding to the peak loads and crack initiation are marked in the curves. The crack initiation in the curve is defined at a scale of around 1 mm. During all the tests, the testing was stopped at an interval of 1 mm after crack initiation, and photographs were taken to measure the crack length along the width direction. The moments when the photographs were taken can also be found in the load–displacement curves, where a decrease of load can be observed due to the suspensions of the testing. The curves of the SEUNT specimens also indicate that cracks initiated after the peak loads are achieved. However, cracks can appear before the peak load is reached for the SEVNT specimens, e.g., SEVNT-10. For the specimens with the same notch depth, the U-notched specimens can sustain larger rupture displacements

Fig. 5.5 Test setup of specimens

(a) Before loading (b) Crack initiation (c) Crack propagation (d) Rupture

Fig. 5.6 Cracking process of SEVNT specimens

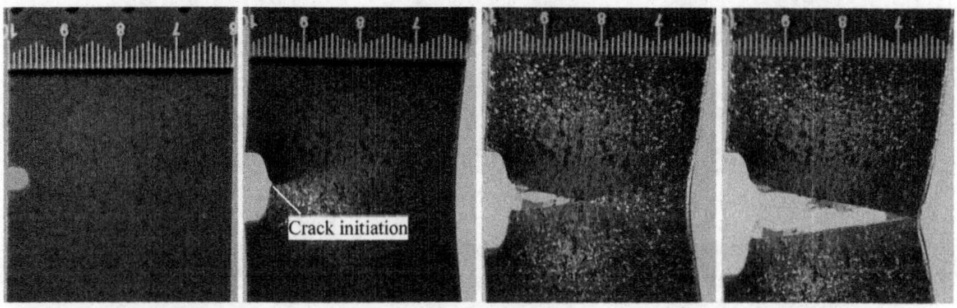

(a) Before loading (b) Crack initiation (c) Crack propagation (d) Rupture

Fig. 5.7 Cracking process of SEUNT specimens

compared with the V-notched ones. As the notch depth increases, the rupture displacement decreases gradually. The test results also give minor differences among the specimens with the same configuration, indicating stable mechanical properties of the ductile fracture specimens and proper setup of the testing. The experimental results listed in Table 5.2 indicate that both the crack initiation displacement and the rupture displacement decrease as the notch depth increases for all the notched specimens. Table 5.2 also shows that the peak loads of the specimens with the same configurations are close to each other, implying stable properties of the steel plate from which the specimens are cut.

The crack propagation curves of the specimens are shown in Fig. 5.10, where the vertical axis is the crack length-to-width ratio and the horizontal one is the displacement. The curves after crack initiation can be simplified as straight lines, indicating an approximately linear relationship between the crack length-to-width ratio and the displacement.

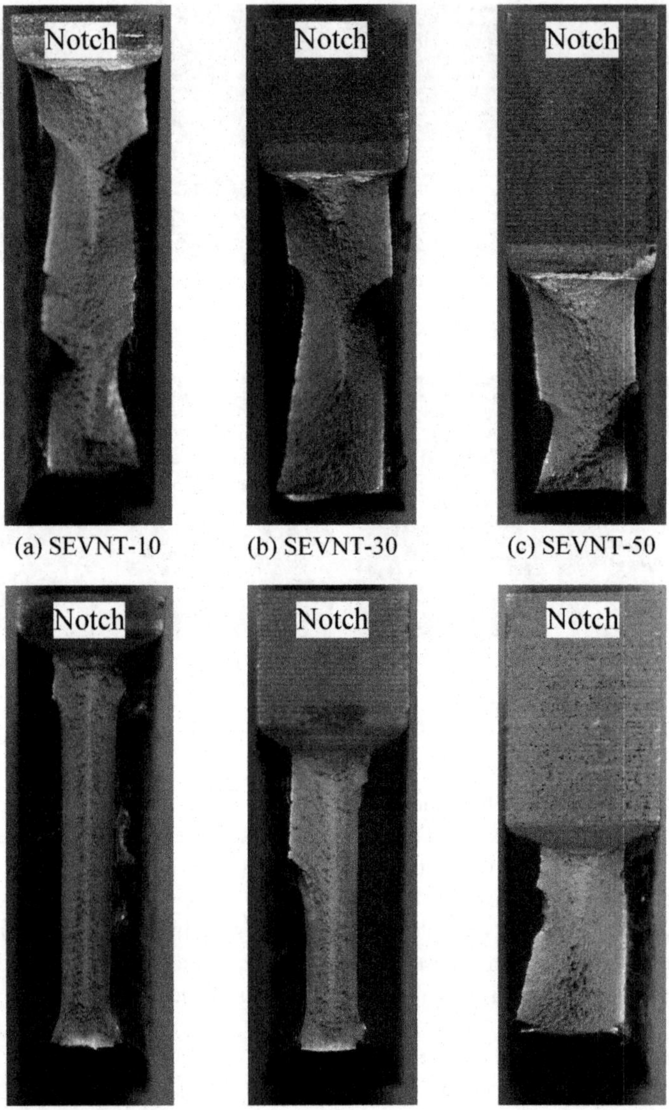

(a) SEVNT-10 (b) SEVNT-30 (c) SEVNT-50

(d) SEUNT-10 (e) SEUNT-30 (f) SEUNT-50

Fig. 5.8 Fracture surfaces of specimens

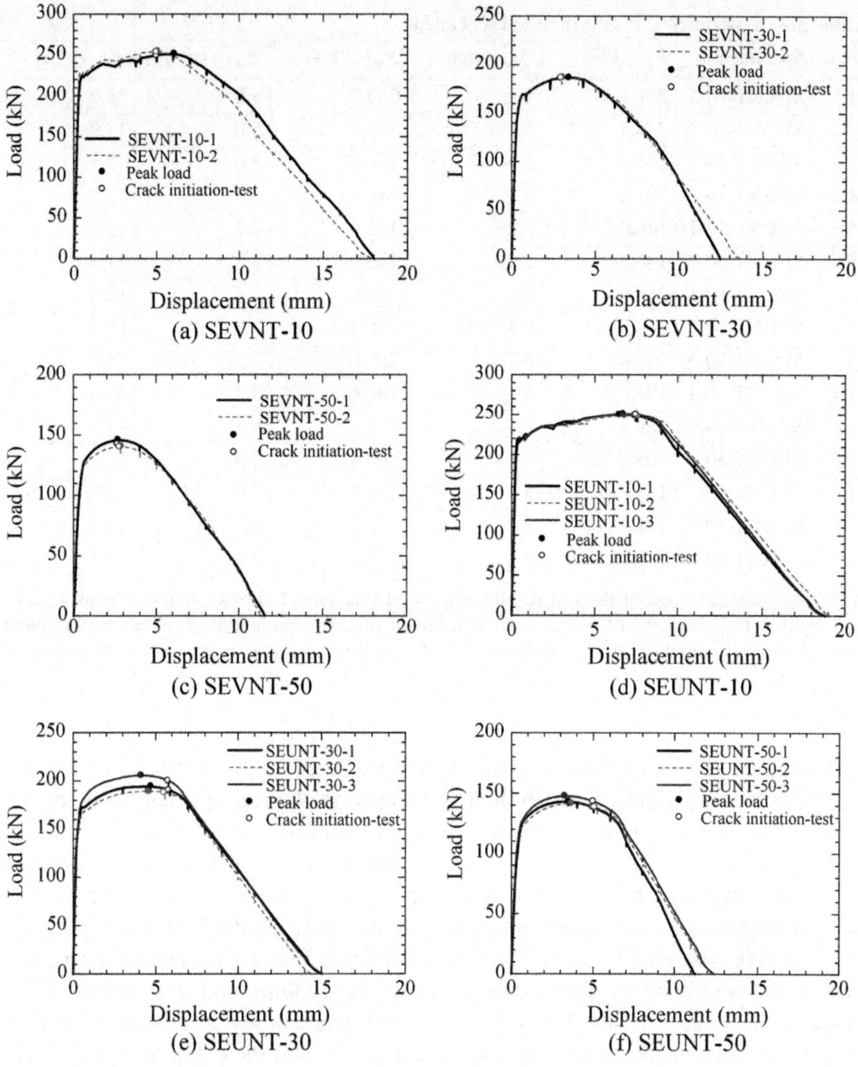

Fig. 5.9 Load–displacement curves of specimens obtained from experimental results

5.4 Numerical Simulation

5.4.1 FE Modeling

Three-dimensional solid models as illustrated in Fig. 5.11 were established to simulate the cracking process of all the specimens, where the actual measured dimensions were employed. Considering the boundary conditions of the experiments, one end

Table 5.2 Experimental results of tested specimens

No.	Specimen	$P_{u,t}$ (kN)	$\Delta_{u,t}$ (mm)	$P_{ini,t}$ (kN)	$\Delta_{ini,t}$ (mm)	$\Delta_{r,t}$ (mm)
1	SEVNT-10-1	251.3	6.1	249.3	5.0	18.0
2	SEVNT-10-2	254.5	5.1	253.5	5.0	17.9
3	SEVNT-30-1	186.8	3.4	186.3	3.0	12.5
4	SEVNT-30-2	187.0	3.1	186.7	3.0	13.8
5	SEVNT-50-1	146.0	2.7	142.4	2.7	11.5
6	SEVNT-50-2	140.7	2.9	140.2	2.8	11.3
7	SEUNT-10-1	250.0	6.5	248.6	7.4	18.8
8	SEUNT-10-2	248.7	6.8	248.1	7.6	19.2
9	SEUNT-10-3	250.8	6.7	250.0	7.5	18.9
10	SEUNT-30-1	194.4	4.7	190.6	5.8	15.0
11	SEUNT-30-2	189.4	4.6	188.0	5.5	14.5
12	SEUNT-30-3	205.7	4.1	200.1	5.8	15.1
13	SEUNT-50-1	143.0	3.5	138.5	5.0	11.3
14	SEUNT-50-2	141.4	3.7	138.3	5.1	12.5
15	SEUNT-50-3	147.6	3.3	143.2	5.0	12.4

Notes $P_{u,t}$ = peak load of the test results; $\Delta_{u,t}$ = displacement corresponding to peak load of test results; $P_{ini,t}$ = load of a specimen at moment of crack initiation; $\Delta_{ini,t}$ = displacement corresponding to crack initiation of test results; $\Delta_{r,t}$ = displacement corresponding to rupture of test results

was fixed and enforced displacement loading was applied to the other end. For the experiments in this study is under monotonic tension, the isotropic hardening model is employed for the plasticity model, where the input true stress–plastic strain data is plotted in Fig. 5.12. Fine mesh was employed at the region close to the notch root to capture the strain concentration accurately, and the other regions were meshed with larger sizes. A convergence study was conducted using the V-notched specimen SEVNT-10, where three mesh schemes as shown in Fig. 5.13, was employed. The typical element sizes of the models with coarse, medium, and fine meshes were, respectively, 3.2, 2.5, and 1.1 mm. The convergence study results are shown in Fig. 5.14, where both the load–displacement curves and the crack length-to-width ratio curves are presented. The results give that the difference between the coarsely meshed model and the other two is relatively large, and the difference between the models with medium and fine mesh is minor. In this study, the fine mesh scheme was employed for all the specimens for the consideration of accuracy. The fracture simulations were conducted using the explicit module in ABAQUS (2010), and the analysis time was selected to ensure the simulations quasi-static. Three-dimensional solid elements with reduced integration (S8R) were employed in the FE models for their good performance in convergence and efficiency for large plastic straining problems.

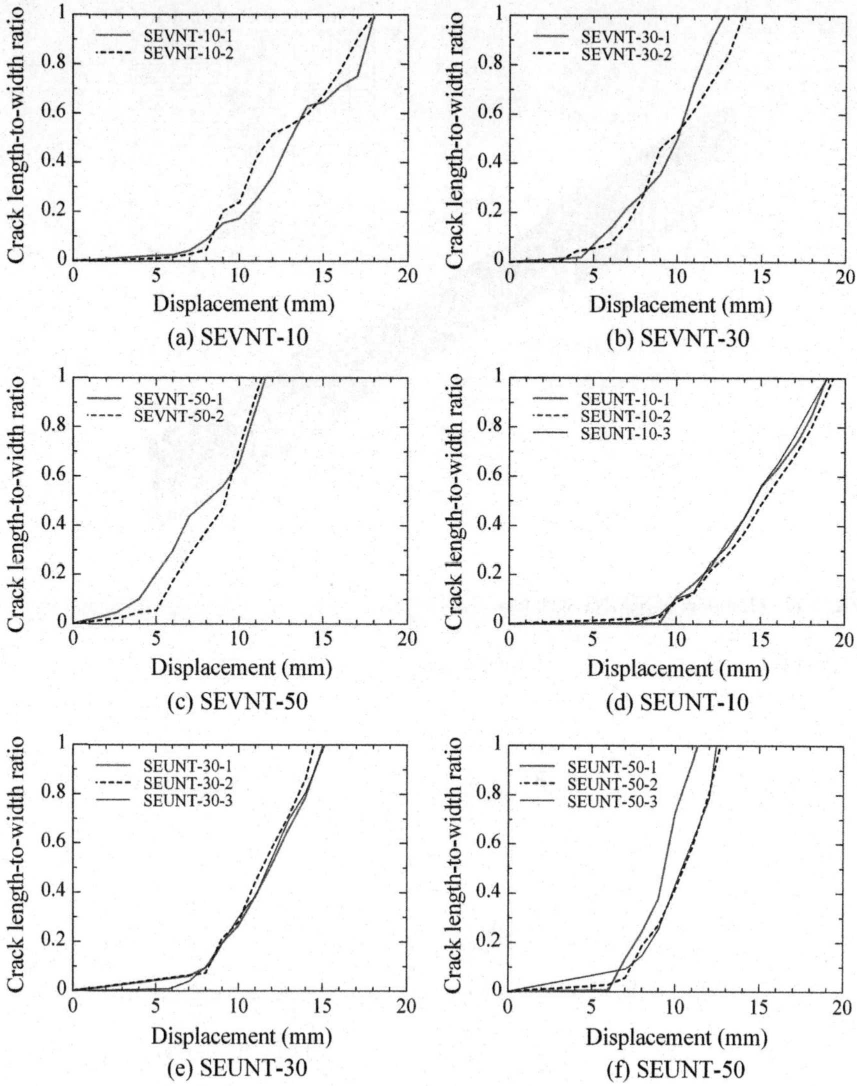

Fig. 5.10 Crack propagation of specimens obtained from experimental results

The fracture model with both crack initiation and crack propagation rules was incorporated in the analyses, where deterioration of stress-carrying capacity was also considered according to Eq. (5.4). The fracture model parameter, χ_{cr}, in Eq. (5.1) can be obtained using the tension coupon tests, which is obtained as 2.4 as listed in Table 5.1. The fracture energy, G_c, in Eq. (5.3) can also be obtained from the tension coupon tests according to the aforementioned calibration method. The average value of G_c using the three coupon test results is 0.22 J/mm^2 as shown in Table 5.1. The

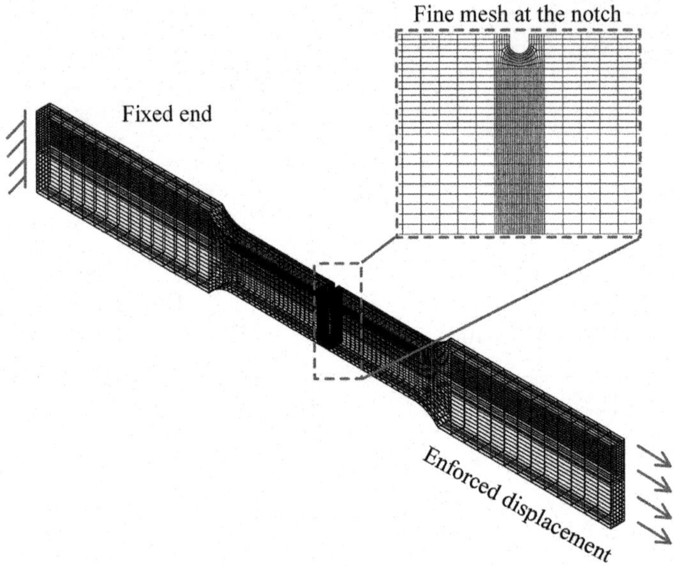

Fig. 5.11 FE model of SEUNT specimen

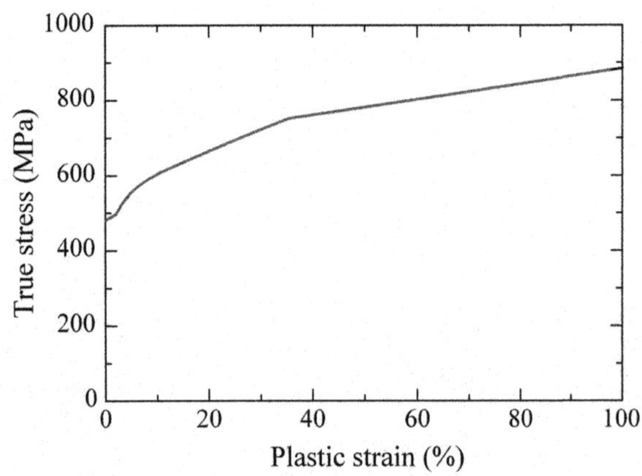

Fig. 5.12 Input true stress–true strain data

value is much smaller than the one obtained in a V-notched Charpy impact coupon in the previous study, and the main reason is that the value calculated using the Charpy impact energy includes a large portion of energy due to plastic straining of the material before crack initiation.

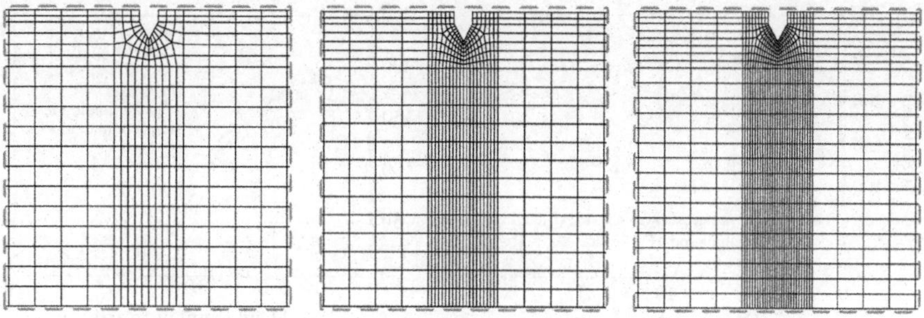

(a) Coarse mesh (b) Medium mesh (c) Fine mesh

Fig. 5.13 Different mesh schemes for SEVNT-10

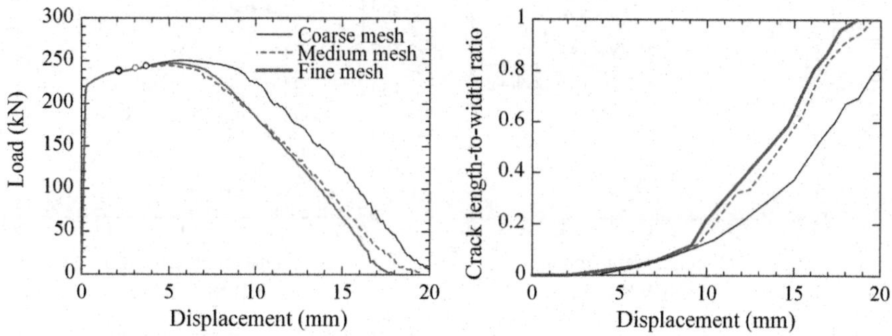

(a) Load-displacement curves and crack initiation (b) Crack propagation

Fig. 5.14 Effect of mesh dependence for cracking simulation of SEVNT-10

5.4.2 *Comparison of Experiment and Simulation*

The experimental load–displacement curves are compared with the corresponding simulation results in Fig. 5.15, where the occurrence of crack initiation is also marked in the curves. The occurrence of crack initiation of the experiments is defined at the same size of the characteristic length of the elements at the notch root. The comparison results indicate that start points of crack initiation of the FE analyses generally compare well for the specimens. The numerical results also can trace the load–displacement curves well as shown in the figure. Comparison of crack propagation between experimental and numerical results is also given in Fig. 5.16, indicating that the fracture model with the proposed parameter calibration method can evaluate the crack propagation process accurately till rupture of the specimens. It should be noted that the ductile fracture model can predict the cracking process till rupture, which is also owing to the fact that the fracture surfaces of all the specimens are 100% ductile. However, for a metal with low fracture toughness, it is possible that there is brittle fracture surface within the whole fractured cross section. The predicted cracking configurations of the SEVNT and SEUNT specimens are, respectively,

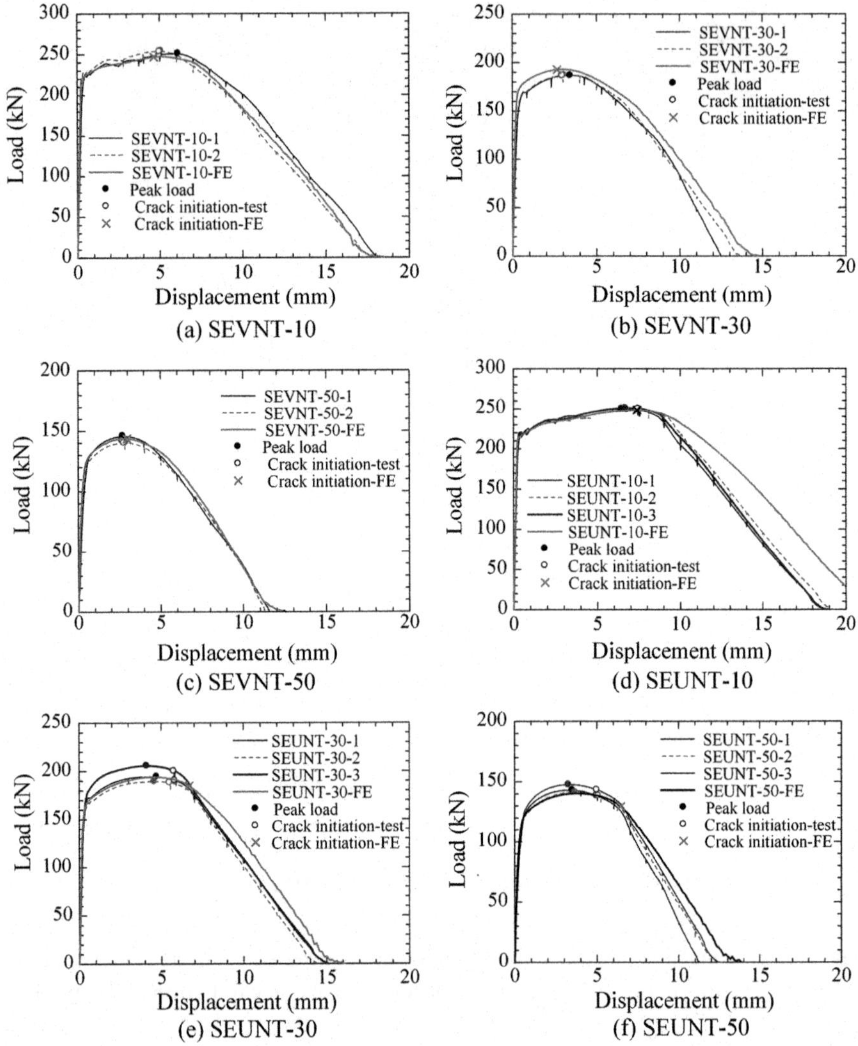

Fig. 5.15 Comparison of load–displacement curves of specimens

illustrated in Figs. 5.17 and 5.18. The same as the experiments, all the cracks initiated at the mid-thickness of the notch roots for all the specimens in the simulations. The cracking configurations are generally well captured, especially for the U-notched specimen, where the cracking path along the horizontal direction is almost a straight line. However, the zigzag fracture paths of the SEVNT specimens cannot be well reproduced in this study due to asymmetric configurations of the initial cracks along the mid-length cross section as illustrated in Fig. 5.6b.

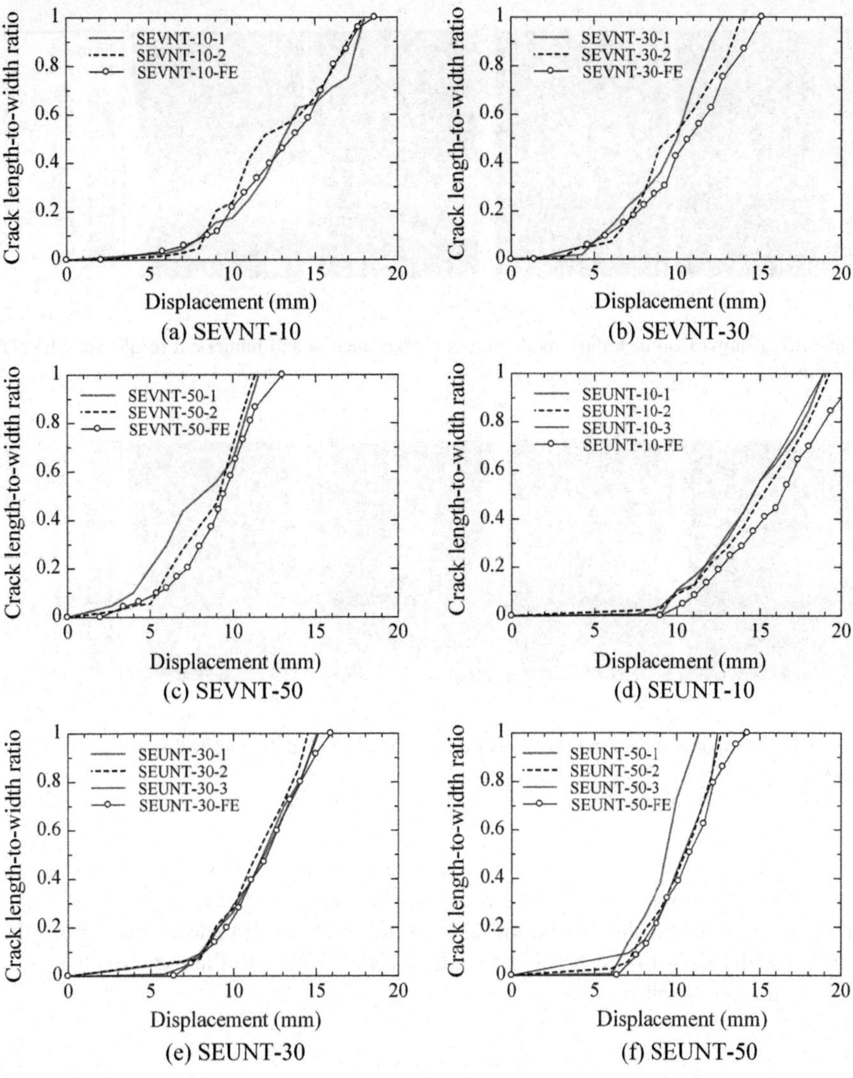

Fig. 5.16 Comparison of crack propagation between experimental and numerical results

Quantitative analyses of the comparison results are given in Table 5.3, including the loads and displacements at different critical states, i.e., peak load, crack initiation, and rupture. Both the loads and the displacements of the experimental results can compare well with the numerical analysis. The averages of the ratios of the numerical results to the corresponding experimental ones range from 0.99 to 1.07, with the coefficients of variation (CoV) ranging from 0.02 to 0.15. It is found that the comparison results of the loads are better than those of the displacements. The CoV of the ratio of the predicted crack initiation displacement to the experimental one,

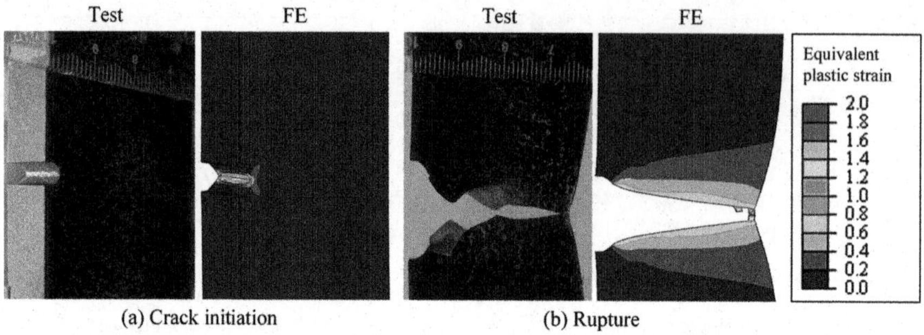

Fig. 5.17 Comparison of failure mode between experimental and numerical results for SEVNT specimens

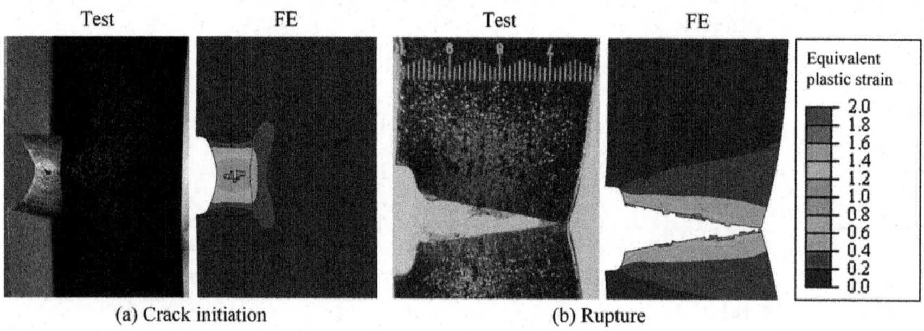

Fig. 5.18 Comparison of failure mode between experimental and numerical results for SEUNT specimens

$\Delta_{\text{ini, FE}}/\Delta_{\text{ini, }t}$, is the largest due to the large measured interval (1 mm) during the experiments, and it is also for the difficulty in accurately measuring the crack length at the mid-thickness during the initial stage of cracking. It can be found from the table that the CoV of the predicted rupture displacement to the experimental one, $\Delta_{r,\text{ FE}}/\Delta_{r,\ t}$, is much smaller.

5.5 Discussion

The cracking process of the smooth tension coupon test was also simulated using the numerical analysis. Figure 5.19 shows the equivalent plastic strain contour plots at different critical states, i.e., before crack initiation, at the crack initiation, and during the crack propagation. Figure 5.19 indicates that the plastic strain range is almost unchanged during crack propagation, indicating that there is almost no plastic straining at the regions except the crack tips during crack propagation. This is mainly owing to the decreasing load-carrying capacity during the rupture process of the smooth tension coupon. The equivalent plastic strain contour plots of the SEUNT

Table 5.3 Comparison of experimental and numerical results

No.	Specimen	$\frac{P_{u,\,FE}}{P_{u,\,t}}$	$\frac{\Delta_{u,\,FE}}{\Delta_{u,\,t}}$	$\frac{P_{\text{ini},\,FE}}{P_{\text{ini},\,t}}$	$\frac{\Delta_{f\ \text{ini},\,FE}}{\Delta_{\text{ini},\,t}}$	$\frac{\Delta_{r,\,FE}}{\Delta_{r,\,t}}$
1	SEVNT-10	0.98	0.91	0.98	0.94	0.99
2	SEVNT-30	1.03	0.94	1.03	0.91	1.09
3	SEVNT-50	1.00	1.00	1.02	1.095	1.05
4	SEUNT-10	0.99	1.10	1.00	0.99	1.17
5	SEUNT-30	0.99	1.00	0.96	1.18	1.03
6	SEUNT-50	0.97	1.11	0.92	1.30	1.10
Average		0.99	1.01	0.99	1.07	1.07
CoV		0.02	0.08	0.04	0.15	0.06

Notes $P_{u,\,FE}$ = peak load of the finite element analysis results; $P_{u,\,t}$ = peak load of the test results; $\Delta_{u,\,FE}$ = displacement corresponding to peak load of finite element analysis results; $\Delta_{u,\,t}$ = displacement corresponding to peak load of test results; $P_{\text{ini},\,FE}$ = load of a specimen at the moment of crack initiation; $P_{\text{ini},\,t}$ = load of a specimen at the moment of crack initiation; $\Delta_{\text{ini},\,FE}$ = displacement corresponding to crack initiation of numerical results; $\Delta_{\text{ini},\,t}$ = displacement corresponding to crack initiation of test results; $\Delta_{r,\,FE}$ = displacement corresponding to rupture of numerical results; $\Delta_{r,\,t}$ = displacement corresponding to rupture of test results

specimens are also given in Fig. 5.20, which indicates that remarkable plastic strain develops in the specimens during the crack propagation stage. It is known that the strain distribution of the Charpy impact coupon is similar to that of the SEUNT specimens. The fracture energy obtained from the Charpy impact coupon includes a large amount of energy to develop plastic deformation, which will greatly overestimate the fracture energy to open a crack. Therefore, the proposed method for the fracture energy calibration using a smooth tension coupon is more accurate since minor plastic strain develops during the crack propagation stage.

5.6 Summaries

A ductile fracture model capable of simulating both ductile crack initiation and crack propagation has been proposed in previous studies, where the crack initiation rule is based on the void growth theory, and the crack propagation rule is in terms of the fracture energy proposed in the literature (Jia et al. 2016). Calibration methods of the fracture parameters related to the two rules are still a main concern for ductile cracking simulation, and commonly specially designed specimens and testing equipment are required, which greatly hinders application of the ductile fracture model. In this chapter, a simple and practical method to obtain the fracture energy, G_c, only using tension coupon tests was proposed. An experimental study was conducted to investigate ductile cracking of single-edge notched tension specimens made of high-strength steel, Q460C, where both V-notched and U-notched specimens were manufactured. A numerical study was also conducted to verify the aforementioned

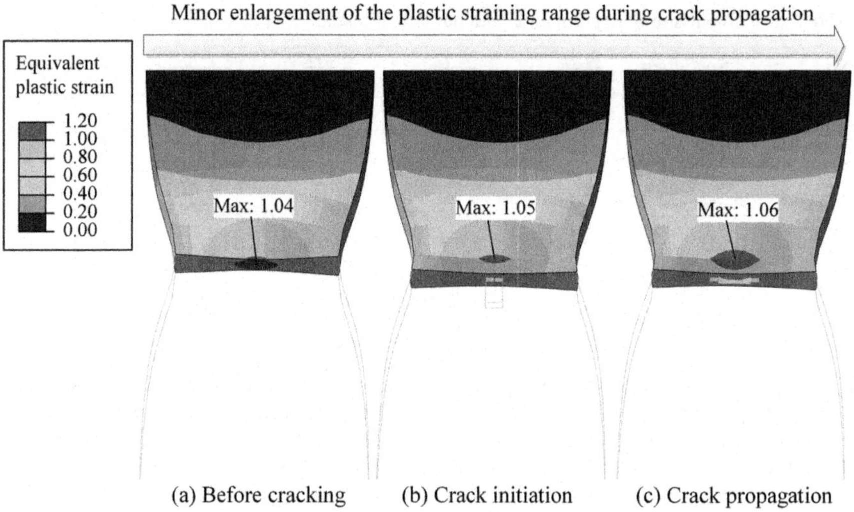

Fig. 5.19 Minor development of plastic strain range during crack propagation of smooth tension coupon

fracture energy calibration method. Based on the experimental and numerical results, the following main conclusions were drawn:

(1) All the fracture surfaces of the steel grade Q460C were 100% ductile, indicating high fracture toughness of the high-strength steel.
(2) The notch root of the V-notched specimens with sharp notch configurations became blunt at the moment of crack initiation, and differences of the rupture displacements between the V- and U-notched specimens were within 10%, indicating that strain concentration had a minor effect on the ductile cracking process of the high toughness steel.
(3) The numerical analyses using the ductile fracture model can accurately evaluate ductile crack initiation and crack propagation till rupture of all the specimens under monotonic tension loading, which validated the applicability of the proposed fracture energy calibration method.
(4) Different notch depths of the V- and U-notched specimens were successfully simulated with the calibrated fracture energy of 0.22 J/mm^2. The average plastic strain levels at the notch roots of the specimens with different notch depths were different from each other. It could thus be concluded that strain constraint had minor effect on the ductile fracture energy.

In this chapter, the proposed ductile fracture energy calibration method has just been validated through monotonic testing on high-strength steel, and more experimental studies are required to investigate other structural metals and also the cases for shear stress dominant loading, and the applicability to cyclic loading should also be further investigated in the future study.

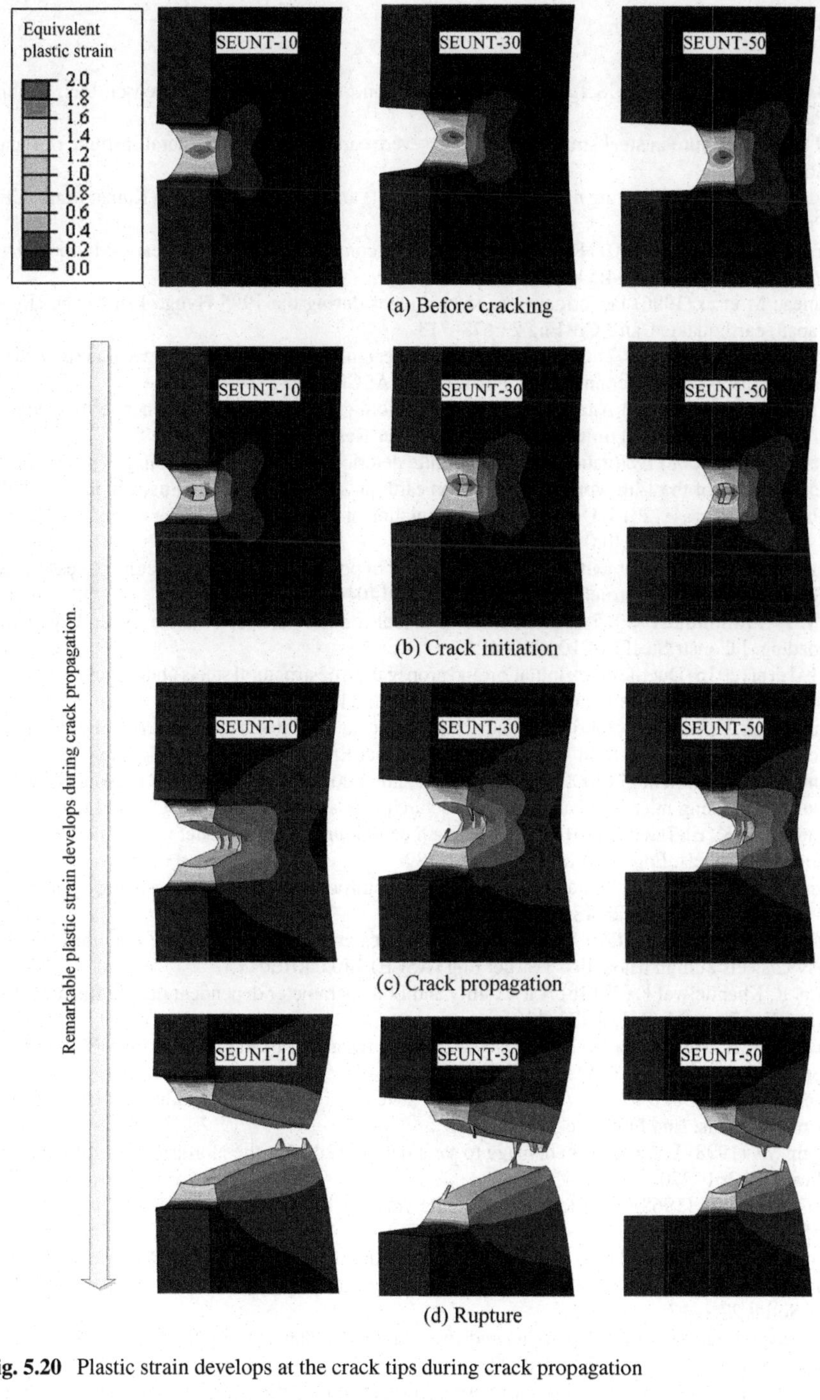

Fig. 5.20 Plastic strain develops at the crack tips during crack propagation

References

ABAQUS (2010) ABAQUS standard manual (version 6.10). Karlsson & Sorensen Inc., Hibbitt, Pawtucket
AIJ (1995) Fracture in steel structures during a severe earthquake. Architectural Institute of Japan, Tokyo
Anderson TL (2005) Fracture mechanics: fundamentals and applications Surjya Kumar Maiti. CRC Press
Barsoum I, Faleskog J (2007) Rupture mechanisms in combined tension and shear—Micromechanics. Int J Solids Struct 44:5481–5498
Bruneau M et al (1996) Performance of steel bridges during the 1995 Hyogo-ken Nanbu (Kobe, Japan) earthquake. Can J Civ Eng 23:678–713
Gilton C, Uang CM (2002) Cyclic response and design recommendations of weak-axis reduced beam section moment connections. J Struct Eng (ASCE) 128:452–463
Hillerborg A et al (1976) Analysis of crack formation and crack growth in concrete by means of fracture mechanics and finite elements. Cem Concr Res 6:773–781
Huang Y et al (2008) Evaluation of steel structure deterioration with cyclic damaged plasticity. In: Proceedings of the 14th world conference on earthquake engineering Beijing, China, 14-0240
Jia L-J, Kuwamura H (2014) Ductile fracture simulation of structural steels under monotonic tension. J Struct Eng (ASCE) 140:04013115
Jia L-J et al (2014) Experimental and numerical study of postbuckling ductile fracture of heat-treated SHS stub columns. J Struct Eng (ASCE) 140:04014044
Jia L-J, Kuwamura H (2015) Ductile fracture model for structural steel under cyclic large strain loading. J Constr Steel Res 106:110–121
Jia L-J et al (2016) Ductile crack initiation and propagation of structural steels under cyclic combined shear and normal stress loading. Constr Build Mater 112:69–83
Kanvinde A, Deierlein G (2006) The void growth model and the stress modified critical strain model to predict ductile fracture in structural steels. J Struct Eng (ASCE) 132:1907–1918
Kanvinde A, Deierlein G (2007) Finite-element simulation of ductile fracture in reduced section pull-plates using micromechanics-based fracture models. J Struct Eng (ASCE) 133:656–664
Khandelwal K, El-Tawil S (2014) A finite strain continuum damage model for simulating ductile fracture in steels. Eng Fract Mech 116:172–189
Kim T et al (2002) Experimental evaluation of plate-reinforced steel moment-resisting connections. J Struct Eng (ASCE) 128:483–491
Kiran R, Khandelwal K (2014a) Experimental studies and models for ductile fracture in ASTM A992 steels at high triaxiality. J Struct Eng (ASCE) 140:04013044
Kiran R, Khandelwal K (2014b) A triaxiality and Lode parameter dependent ductile fracture criterion. Eng Fract Mech 128:121–138
Kuwamura H, Yamamoto K (1997) Ductile crack as trigger of brittle fracture in steel. Journal of Structural Engineering (ASCE) 123:729–735
Mackenzie AC et al (1977) On the influence of state of stress on ductile failure initiation in high strength steels. Eng Fract Mech 9:167–188
Mahin SA (1998) Lessons from damage to steel buildings during the Northridge earthquake. Eng Struct 20:261–270
McClintock FA (1968) A criterion for ductile fracture by the growth of holes. J Appl Mech 35:363–371
Miner MA (1945) Cumulative damage in fatigue. J Appl Mech 12:A159–A164
Nahshon K, Hutchinson JW (2008) Modification of the Gurson model for shear failure. Eur J Mech A Solids 27:1–17
O'Sullivan D et al (1998) Repairs to mid-rise steel frame damaged in Northridge earthquake. J Perform Constructed Facil 12:213–220
Panontin TL, Sheppard SD (1995) The relationship between constraint and ductile fracture initiation as defined by micromechanical analyses. ASTM STP 1256, ASTM, West Conshohoken, pp 54–85

Qian X, Yang W (2012) Initiation of ductile fracture in mixed-mode I and II aluminum alloy specimens. Eng Fract Mech 93:189–203

Rice JR, Tracey DM (1969) On the ductile enlargement of voids in triaxial stress fields. J Mech Phys Solids 17:201–217

Roufegarinejad A, Tremblay R (2012) Finite element modeling of the inelastic cyclic response and fracture life of square tubular steel bracing members subjected to seismic inelastic loading. In: Mazzolani F, Herrera R (eds) STESSA 2012, Taylor and Francis Group, pp 97–103

Rousselier G (1987) Ductile fracture models and their potential in local approach of fracture. Nucl Eng Des 105:97–111

Sumner E, Murray T (2002) Behavior of extended end-plate moment connections subject to cyclic loading. J Struct Eng (ASCE) 128:501–508

Tvergaard V (2009) Behaviour of voids in a shear field. Int J Fract 158:41–49

Tvergaard V, Nielsen KL (2010) Relations between a micro-mechanical model and a damage model for ductile failure in shear. J Mech Phys Solids 58:1243–1252

Zhou H et al (2013) Extremely low cycle fatigue prediction of steel beam-to-column connection by using a micro-mechanics based fracture model. Int J Fatigue 48:90–100

Chapter 6
Ductile Fracture Model of Structural Steel under Ultra-low-Cycle Fatigue Loading

Abstract For specimens under ultra-low-cycle fatigue loading, cyclic plasticity and negative stress triaxialities are two main differences from monotonic loading cases. In this chapter, a crack initiation rule by adding a damage accumulation rule under negative stress triaxialities is proposed, where no damage accumulation is assumed for stress triaxialities below −1/3. This rule is based on previous experimental results. This chapter is mainly focused on the crack initiation rule of ductile metal under ultra-low-cycle fatigue loading (ULCF). Likewise, crack propagation is assumed to occur once a crack initiates. The cyclic ductile fracture model is validated through comparison with cyclic tests on hourglass-type specimens under a series of different loading histories. Numerical fracture simulation using both the previously proposed cyclic plasticity models and the cyclic ductile fracture model is conducted, and the analysis can well predict ductile cracking of the specimens with relatively uniform stress and strain distribution.

6.1 Introduction

Ductile fracture under seismic loading is often mistaken as low-cycle fatigue, since the two fracture modes are both resulted by cyclic loading in a small number of cycles. Kuwamura (1997) found that there is a transition of fracture modes from low-cycle fatigue to ductile fracture depending on the elevation of strain amplitude based on cyclic loading test results on hourglass-type notched coupons. During an earthquake event, structures commonly suffered from large plastic loading within hundreds of cycles, and the fracture mode should be classified as ductile fracture but not low-cycle fatigue. The aforementioned findings are of great importance for the study of fracture of steel structures under seismic loading, since the mechanisms and evaluation approaches of the two fracture modes are distinguished. Ductile fracture has a typical dimple fracture surface under fractographic observations with a scanning electron microscope (SEM), and it is possible to predict ductile fracture with only small-scale monotonic tension coupon tests. However, typical SEM fractography of low-cycle fatigue is striations. It may be difficult to predict low-cycle fatigue

L.-J. Jia and H. Ge, *Ultra-low-Cycle Fatigue Failure of Metal Structures under Strong Earthquakes*, Springer Tracts in Civil Engineering,
https://doi.org/10.1007/978-981-13-2661-5_6

life from a monotonic tension coupon tests. To distinguish the loading history of ductile fracture from that of low-cycle fatigue which is concerned with small strain amplitudes, the loading history concerned with ductile fracture under seismic loading is termed as ultra-low-cycle fatigue (ULCF) loading in this book.

Bonora (1997) proposed a continuum damage mechanics (CDM) model with five model parameters for ductile fracture under monotonic loading, where the calibration of the parameters requires performing a number of unloading tests using a smoothly-notched coupon, which is not convenient to apply to structural engineering. Pirondi and Bonora (2003) extended the CDM model to predict ductile fracture under ULCF loading, where the damage index, D, is accumulated if and only if the stress triaxiality is positive. The damage rule and mechanism at negative stress triaxiality are still open to discussion, but the assumption that no damage accumulates at negative stress triaxialities is not realistic based on former test results, e.g., (Bao and Treitler 2004; Enami 2005).

Research on ductile fracture of metals under ULCF loading is found in a few literatures. A fracture model based on the concept of effective stress is proposed by Ohata and Toyoda (2004), and fracture was assumed to occur when the isotropic hardening (IH) stress reaches a threshold value. The model was also applied to ductile fracture prediction of circular tubes under cyclic incremental bending (Toyoda 2000) and beam-to-column connections of steel piers under ULCF loading (Yasuda et al. 2004). However, this model may not be applicable to constant-amplitude cyclic loading, since the IH component will stabilize quickly and the material will not fracture according to the damage rule, which is contrary to real situation.

Bao and Treitler (2004) proposed an empirical fracture model with three model parameters based on tension tests on smoothly-notched aluminum coupons with various pre-compressive strains. The formation of the model is simple, while a number of cyclic tests are necessary to obtain the relationship between the stress triaxiality and the fracture strain. Bai et al. (2006) also proposed an empirical one-parameter fracture model for ductile fracture with various pre-compressive strains, where fracture was postulated to occur when the stress modified equivalent plastic strain reaches a critical value. The model was also employed to study ductile fracture of aluminum stub columns under compressive loading, where reversal straining is concerned. The model is inconvenient to apply to structural engineering since a compression and tension test on smoothly-notched coupon is required to obtain the model parameter. Moreover, the accuracy of the model for ductile fracture of structural steels under ULCF loading requires further study since the model was calibrated through tests on aluminum.

Study on ductile fracture of structural steels under ULCF loading is more limited. A semi-empirical model based on the Rice–Tracey model was proposed by Kanvinde and Deierlein (2007) to predict ductile fracture of structural steels under ULCF loading, and the model was also applied to ductile fracture prediction of bluntly-notched and dogbone steel specimens (Kanvinde and Deierlein 2008). However, the definitions of the two fracture parameters of the CVGM model are quite complicated, and the parameters have to be obtained from monotonic and cyclic tests of the smoothly-notched coupons, which may be difficult and even impossible for structural engineers due to limitations of testing equipment and configurations of structural members. No specific approach to obtain the model parameters of the plasticity model was given. Moreover, modification of the true stress–true strain has not been carried out during the identification of the model parameters, which may result in some deviation during numerical simulation of the smoothly-notched coupons due to the non-uniform distribution of stress over the minimum cross section.

In this chapter, the monotonic fracture model proposed in Chap. 4 (Jia and Kuwamura 2014) is extended to predict ductile fracture of structural steels under ULCF loading (Jia and Kuwamura 2015), where the monotonic fracture model is modified by considering proper damage rule under negative stress triaxialities, and the damage rule of the cyclic fracture model is consistent with the monotonic damage model for positive stress triaxialities. The prediction of ductile fracture under cyclic loading is more complicated than that of monotonic tension. The difficulties may involve three aspects, i.e., a fracture model for cyclic compression and tension; cyclic plasticity model to evaluate cyclic stress–strain behaviors; and straightforward methods to calibrate the corresponding model parameters related with the fracture model and the plasticity model. It was found that steels will not fracture when the stress triaxiality is less than a cut-off value of $-1/3$ (Bao and Wierzbicki 2005). Therefore, the fracture model is slightly modified by assuming that the damage index, D, will not evolve when the stress triaxiality is less than $-1/3$. The one-parameter cyclic fracture model proposed in this study can predict ductile fracture under both monotonic and cyclic loading, where the only one fracture parameter can be obtained from a tension test on a common smooth coupon straightforwardly.

The cyclic tests on hourglass-type coupons under various loading histories are adopted herein to validate the proposed cyclic fracture model. Numerical simulation of the ductile fracture of the experiments is also carried out, where the two cyclic plasticity models validated in Chap. 3, i.e., the Chaboche model with IH and the modified Yoshida–Uemori model, are adopted to simulate the cyclic plasticity behaviors of the material. Finally, the numerical simulation results are compared with the experimental ones, from which the cyclic fracture model is proved to be capable of simulating ductile fracture of structural steels under ULCF loading with favorable accuracy.

6.2 Ductile Fracture Model under Ultra-low-Cycle Fatigue Loading

6.2.1 Damage of Metals under Negative Stress Triaxiality

In structural engineering, ductile fracture of structural members may occur within a small number of cycles during a seismic event. Ductile fracture of steels under ULCF loading is quite different from that under monotonic tension, since negative stress triaxialities are involved at the cracked regions for cyclic loading, while the stress triaxialities are positive for the case of monotonic tension as investigated in the preceding chapter.

A large number of tension tests on various steels and stainless steels using round bars have been carried out under different negative stress triaxialities by Bridgman (1952), where the coupons were tested under surrounding hydrostatic pressures to produce negative stress triaxialities. It has been found that the fracture strain increases as the superimposed hydrostatic pressure increases. A series of tension fracture tests with hydrostatic pressure on different metals were carried out by French and Weinrich (1973, 1975a, b), French et al. (1973). Axial compression tests using short cylinders were also carried out to study ductile fracture of metals under negative stress triaxialities, e.g., (Bao and Wierzbicki 2004; Thomason 1969). Tension tests of two steels with different compressive prestrain were carried out by Enami (2005), which also indicates the ductility doesn't decrease up to compressive prestrain of 30% for one steel and 10% for the other one. French and Weinrich (1975a) has pointed out the fracture strain of copper approaches infinity when the superimposed hydrostatic pressure is beyond a certain value. Bao and Wierzbicki (2005) has derived a cut-off value of negative stress triaxiality of $-1/3$, below which no fracture will occur, based on the test results of Bao and Wierzbicki (2004), Bridgman (1952), Kudo and Aoi (1967).

6.2.2 *Modification of the Monotonic Fracture Model to Apply to Cyclic Large Strain Loading*

On the basis of above discussions, it is assumed in this study that no damage will accumulate when the stress triaxiality is below $-1/3$. Then, the evolution rule of the fracture model in Eq. (4.7) can be modified for cyclic loading (Jia and Kuwamura 2015),

$$\mathrm{d}D = \begin{cases} \dfrac{\mathrm{d}\varepsilon_{\mathrm{eq}}^{p}}{\chi \cdot \mathrm{e}^{-\frac{3}{2}T}} & \text{when } T \geq -\frac{1}{3} \\ 0 & \text{when } T < -\frac{1}{3} \end{cases} \tag{6.1}$$

The fracture strain can be given as illustrated in Fig. 6.1, where the fracture strain in the stress triaxiality space is the same as that of the Rice–Tracey model when T is larger than $-1/3$, and equal to infinity when T is below $-1/3$. The fracture model proposed for ductile fracture under monotonic loading in Chap. 4 can be easily modified for cyclic loading.

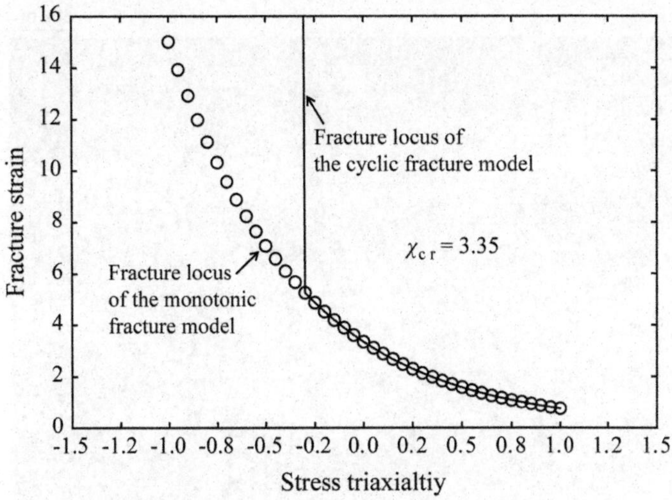

Fig. 6.1 Fracture strain versus stress triaxiality of fracture model ($\chi = 3.35$)

Fig. 6.2 Fracture surface observed by a common camera (*KA05*)

6.3 Experiment

As is mentioned in Chap. 3, the hourglass-type specimens for cyclic loading (except *KA02*) were all pulled to fracture after being subjected to different ULCF loading histories. The test results are herein employed to calibrate the fracture model under ULCF loading. The details of the tests can refer to Chap. 3. All the fracture surfaces of the specimens are the typical "cup and cone" appearance as illustrated in Fig. 6.2, and the dimple pattern as illustrated in Fig. 6.3 can be observed using an SEM, indicating a ductile fracture mode.

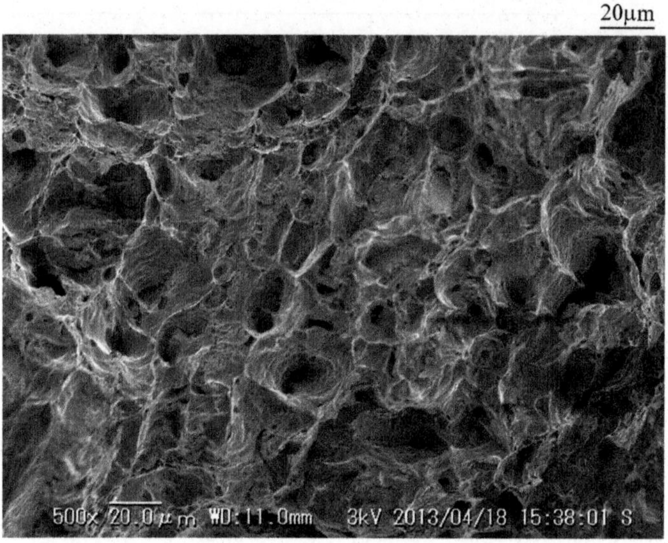

Fig. 6.3 Fractographic observation result of structural steel using SEM (*KA05*)

6.4 Numerical Simulation

6.4.1 FE Modeling

Numerical simulation was carried out to simulate ductile fracture of the aforementioned test results of the hourglass-type specimens under ULCF loading using the explicit module in ABAQUS (2010). The explicit module in ABAQUS is an FE analyzer that employs explicit integration scheme to solve highly nonlinear systems, e.g. complicated problems with related with contact, dynamic, and failure events. Herein, quasi-static analyses were carried out using the explicit module to simulate the ductile fracture of the experiments under ULCF loading. Long enough step time was assigned to each model to ensure quasi-static analysis where the corresponding inertial force is relatively small enough compared with the internal energy of each model. Axisymmetric half models using CAX8 elements were adopted, and both whole models and part models as shown in Fig. 6.4 were built. The part models were utilized for the analyses since it was found the two models gives almost the same load–displacement curves.

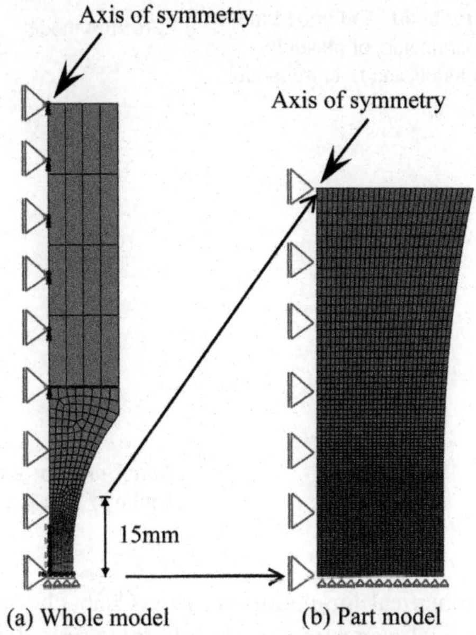

Fig. 6.4 Mesh of FE models

6.4.2 *Plasticity Model*

It has been validated in Chap. 3 that the Chaboche model with IH and the modified Yoshida–Uemori model can well evaluate the stress–strain behaviors under cyclic large strain loading, and the model parameters can be calibrated using only monotonic tension coupons. Herein, the tests were simulated using the two plasticity models, i.e., the Chaboche model with IH and the modified Yoshida–Uemori model, combined with the proposed fracture model, respectively. All the model parameters of the plasticity models and the fracture model were calibrated using only the tension coupon test results. The true stress–true strain data was first obtained using the MWA method (Jia and Kuwamura 2014) proposed in Chap. 2, and the model parameters were obtained through the corresponding method given in Chap. 3. The calibrated model parameters for the two plasticity models and the fracture model are as listed in Table 6.1.

6.5 Comparison of Experiment and Simulation

The comparison results are shown in Fig. 6.5, where the result of *KA02* is not shown due to premature buckling of the specimen before fracture. The numerical simulation results using the two cyclic plasticity models and the fracture model compare well with the corresponding experimental results both for the monotonic test (*KA01*) and ULCF loading tests (*KA03–KA07*). For *KA04* with a constant-amplitude loading, the

Table 6.1 Calibrated model parameters of plasticity models and fracture model

Fracture model		Chaboche models with isotropic hardening		Modified Yoshida–Uemori model	
χ	3.35	σ_{y0}	255.9	σ_{y0}	255.9
		C_1	26.9	C	332.8
		C_2	26.9	B	321.7
		C_3	1617.2	R_{sat}	137.7
		γ_1	0	b	82.9
		γ_2	0	m	18.1
		γ_3	10.7	h	0.5
		k	5.8	m_l	236.2
		Q_∞	227.8	$\varepsilon_{plateau}$	0.0148

Note Unit of σ_{y0}, C_1, C_2, C_3, Q_∞, C, B, R_{sat}, b, m_l: MPa
Unit of χ, γ_1, γ_2, γ_3 and k, m, $\varepsilon_{plateau}$: dimensionless

numerical simulation using the Chaboche model with IH combined with the fracture model overestimates slightly the stress after the second constant-amplitude cycle, while it globally gives almost the same fracture point with that using the modified Yoshida–Uemori model. The overestimation is due to the lack of a memory surface in the Chaboche model with IH, while the modified Yoshida–Uemori model has a memory surface in strain space to memorize the loading history, which can simulate the non-IH hardening effect under constant-amplitude loading.

The comparison between the experimental results of *KA05* and *KA06* indicates that the damage under strain ranges close to fracture strain of the material is much larger than that at small strain ranges, e.g., before necking. The numerical simulations using the two plasticity models and the proposed fracture model gives good evaluation of the fracture point of *KA05*, while overestimate the ductility of *KA06*. The overestimation may be due to that the proposed fracture model underestimates the damage within the strain ranges close to fracture strain under negative stress triaxialities.

6.6 Summaries

This chapter provides a straightforward method to simulate ductile fracture of structural steels under ULCF loading. A fracture model for ULCF loading with only one parameter has been proposed. Experimental tests on a mild steel under various loading histories were carried out using hourglass-type specimens. Numerical simulations were also carried out to predict ductile fracture under ULCF loading using the combinations of the plasticity models and the proposed fracture model. The following conclusions can be drawn

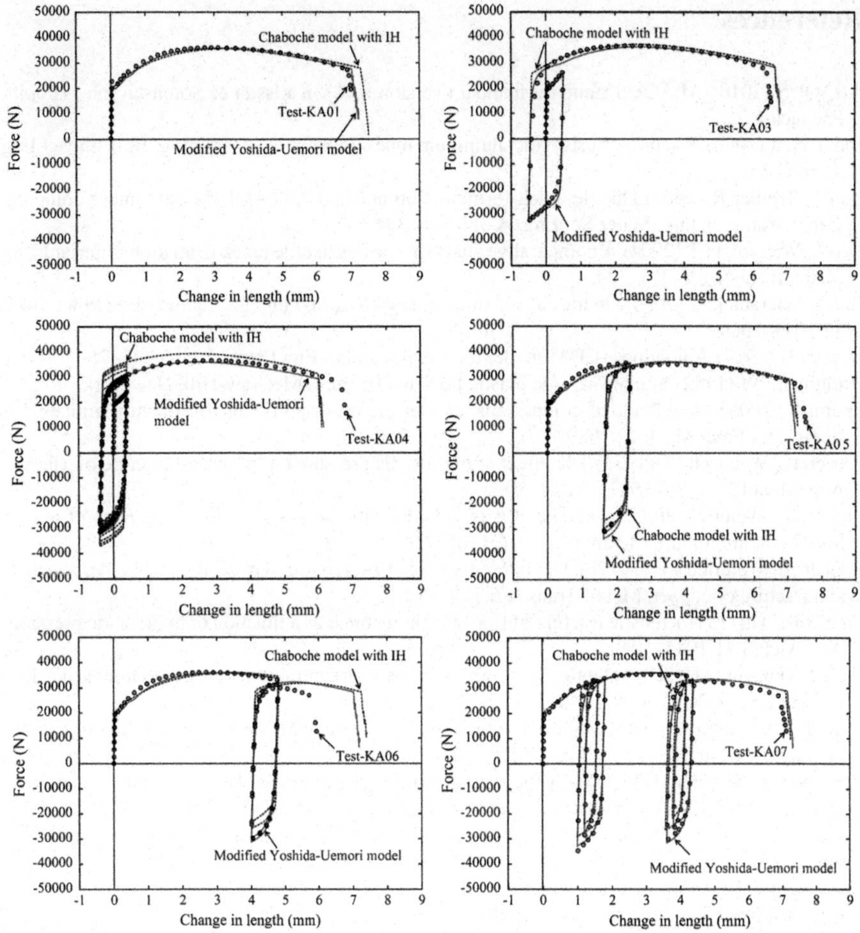

Fig. 6.5 Comparison between test and numerical results

(1) A simple fracture model was proposed, which combined the Rice–Tracey model and the Miner's rule in incremental form, and assumed that there is no damage under stress triaxialities less than a cut-off value. The fracture model was generally applicable to structural steels under both monotonic and ULCF loading tests.
(2) The numerical simulations using the combinations of the two plasticity models, i.e., the Chaboche model with IH and the modified Yoshida–Uemori model, and the proposed fracture model can well simulate the ductile fracture initiation under ULCF loading.
(3) The Chaboche model with IH overestimates slightly the stress under constant-amplitude loading, while the predicted fracture point is almost the same with that of the modified Yoshida–Uemori model.

References

ABAQUS (2010) ABAQUS standard manual (Version 6.10). Karlsson & Sorensen Inc., Hibbitt. Pawtucket

Bai Y et al (2006) Fracture of prismatic aluminum tubes under reverse straining. Int J Impact Eng 32:671–701

Bao Y, Treitler R (2004) Ductile crack formation on notched Al2024-T351 bars under compression–tension loading. Mater Sci Eng, A 384:385–394

Bao Y, Wierzbicki T (2004) A comparative study on various ductile crack formation criteria. J Eng Mater Technol 126:314–324

Bao Y, Wierzbicki T (2005) On the cut-off value of negative triaxiality for fracture. Eng Fract Mech 72:1049–1069

Bonora N (1997) A nonlinear CDM model for ductile failure. Eng Fract Mech 58:11–28

Bridgman PW (1952) Studies in large plastic flow and fracture. McGraw-Hill, New York

Enami K (2005) The effects of compressive and tensile prestrain on ductile fracture initiation in steels. Eng Fract Mech 72:1089–1105

French IE, Weinrich PF (1973) The effect of hydrostatic pressure on the tensile fracture of α-brass. Acta Metall 21:1533–1537

French IE, Weinrich PF (1975a) The effects of hydrostatic pressure on the mechanism of tensile fracture of aluminum. Metall Trans A 6:1165–1169

French IE, Weinrich PF (1975b) The influence of hydrostatic pressure on the tensile deformation and fracture of copper. Metall Trans A 6:785

French IE et al (1973) Tensile fracture of free machining brass as a function of hydrostatic pressure. Acta Metall 21:1045–1049

Jia L-J, Kuwamura H (2014) Ductile fracture simulation of structural steels under monotonic tension. J Struct Eng (ASCE) 5(140):04013115

Jia L-J, Kuwamura H (2015) Ductile fracture model for structural steel under cyclic large strain loading. J Constr Steel Res 106:110–121

Kanvinde A, Deierlein G (2007) Cyclic void growth model to assess ductile fracture initiation in structural steels due to ultra low cycle fatigue. J Eng Mech (ASCE) 133:701–712

Kanvinde A, Deierlein G (2008) Validation of cyclic void growth model for fracture initiation in blunt notch and dogbone steel specimens. J Struct Eng (ASCE) 134:1528–1537

Kudo H, Aoi K (1967) Effect of compression test conditions upon fracturing of medium carbon steel. J Jpn Soc Tech Plasticity 8:17–27

Kuwamura H (1997) Transition between fatigue and ductile fracture in steel. J Struct Eng (ASCE) 123:864–870

Ohata M, Toyoda M (2004) Damage concept for evaluating ductile cracking of steel structure subjected to large-scale cyclic straining. Sci Technol Adv Mater 5:241–249

Pirondi A, Bonora N (2003) Modeling ductile damage under fully reversed cycling. Comput Mater Sci 26:129–141

Thomason PF (1969) Tensile plastic instability and ductile fracture criteria in uniaxial compression tests. Int J Mech Sci 11:187–198

Toyoda M (2000) Ductile fracture initiation behavior or pipe under a large scale of cyclic bending. Pipeline Technol 2

Yasuda O et al (2004) Application to Ductile cracking evaluation for beam-to-column connection of Steel Pier: evaluation of Ductile crack initiation for welded structures subjected to large scale cyclic loading (report 2). Q J Jpn Weld Soc 22:467–476

Chapter 7
Application to Post-buckling Ductile Fracture of Steel Stub Columns

Abstract In practice, ductile fracture after occurrence of local buckling has been observed in actual steel members during strong earthquakes, such as the 1995 Kobe earthquake. This chapter aims to establish a simulation method for the post-buckling cracking process on the basis of formerly proposed cyclic fracture and plasticity models. Experimental tests on twelve heat-treated square hollow section stub columns are carried out under different cyclic loading histories. Numerical simulation of the post-buckling ductile cracking of the specimens is conducted, where the simulation results indicate that the crack initiation of the specimens can be evaluated with a favorable accuracy, while further study is required for the crack propagation under ultra-low-cycle fatigue loading.

7.1 Introduction

During a strong earthquake event, local or global buckling may first occur in some crucial steel structural members sustaining large deformation such as bracings and corner columns, and ductile fracture may follow them during the rest of large displacement loading cycles, as observed in the reports of the 1994 Northridge earthquake and the 1995 Kobe earthquake. This chapter aims to investigate the cases that ductile fracture occurs after local buckling under cyclic quasi-static loading.

In this chapter, a series of steel square hollow section (SHS) stub columns were designed and experimentally tested. Two factors, i.e., the width-to-thickness ratio and loading history, were investigated in the experiments. Twelve heat-treated specimens were manufactured from three long cold-formed box columns with various thicknesses. Three width-to-thickness ratios and four different loading protocols were studied. Tension coupons cut from three various locations of the columns, i.e., corner part, parallel part, and weld part, were also manufactured to obtain the material properties of the whole columns. Numerical simulation using the plasticity model (Jia and Kuwamura 2014b) and cyclic fracture model (Jia and Kuwamura 2015) was also carried out to investigate the failure mechanisms of the experiments. The numerical simulation can predict crack initiation of the experiments with acceptable accuracy.

L.-J. Jia and H. Ge, *Ultra-low-Cycle Fatigue Failure of Metal Structures under Strong Earthquakes*, Springer Tracts in Civil Engineering,
https://doi.org/10.1007/978-981-13-2661-5_7

7.2 Experiment

7.2.1 Specimen

Since the width-to-thickness ratio of an SHS column is a dominant factor for local buckling, and seismic waves have a large variety of types, the effects of the two variables are investigated. In this chapter, heat-treated cold-formed SHS stub columns with three different width-to-thickness ratios were loaded under four different loading histories.

Columns as illustrated in Fig. 7.1 were tested under quasi-static loading, where the columns are made of STKR400. The measured geometrical parameters of the specimens are listed in Table 7.1. The length and width of all the specimens were designed as 300 and 100 mm, respectively, and two circular 30-mm-thick loading plates were welded to the columns using full penetration welds. Backing plates were also employed to assure full penetration of the welds between the columns and the loading plates. The full penetration welds are of great importance to prevent the premature failure of the welded connections. The weld beads of the columns were removed before the columns were welded to the loading plates. The four surfaces of the columns are numbered, where the surface with a seam weld is numbered as Surface “1”.

The columns were heat-treated to 870 °C and held at this temperature for an hour before they were connected to the loading plates.

Since the material properties of the SHS columns may differ over the cross section due to the forming process, tension coupons were cut from the parrallel, the corner and the welding parts, to obtain the average properties of the section. For the welding part, the inner weld beads of the columns were removed before manufacturing of the coupons. For each type of the coupons, three coupons were made to study the material deviations and to obtain the average material properties. The mechanical properties and chemical composition are listed in Table 7.2. The average material

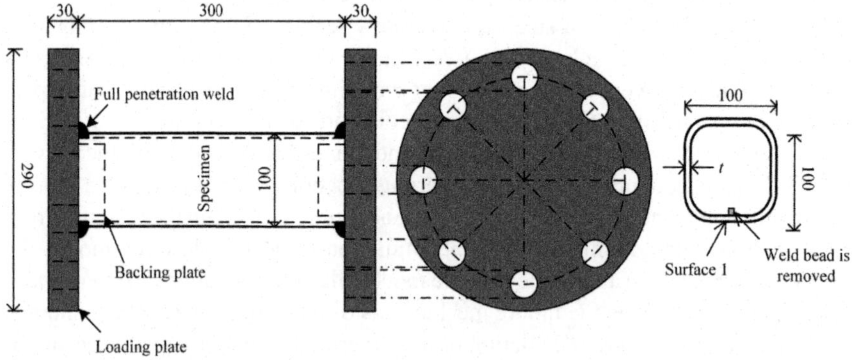

Fig. 7.1 Configuration of test specimens

Table 7.1 Measured geometrical and mechanical properties of specimens

Specimens	t (mm)	Width-to-thickness ratio	σ_{y0} (N/mm^2)	σ_{buckle}^{ini} (N/mm^2)	σ_u^c (N/mm^2)	Loading history
RH1-1	2.1	47.6	258.7	–	–	Monotonic tension
RH1-2[a]			–	253.2	267.9	Two cycles[a]
RH1-3			260.0	203.1	203.1	Incremental
RH1-4[a]			–	252.6	252.6	Constant amplitude[a]
RH2-1	4.2	23.8	270.7	–	–	Monotonic tension
RH2-2			227.0	274.0	274.0	A single full cycle
RH2-3			242.3	253.6	253.6	Incremental
RH2-4			217.0	274.1	274.1	Constant-amplitude
RH3-1	8.4	11.9	242.4	–	–	Monotonic tension
RH3-2			225.5	304.4	304.4	A single full cycle
RH3-3			209.8	295.1	362.1	Incremental
RH3-4			243.6	305.1	362.9	Constant-amplitude

Notes σ_{y0} is the average initial tensile yield stress; σ_{buckle}^{ini} and σ_u^c denote the average initial buckling stress and maximum average compressive stress, respectively

[a]To design proper loading history for the tests, specimens RH1-2 and RH1-4 were pilot tested under different loading histories. Specimen RH1-2 was loaded under two cycles, followed by pull to fracture, and RH1-4 was under constant-amplitude loading, while the initial half cycle is compressive loading but not tensile loading

properties at different locations of the heat-treated columns with the same thickness are close to each other.

7.2.2 *Test Setup*

The experiments were carried out using a universal testing machine with a maximum tensile capacity of 2000 kN and a maximum compressive capacity of 5000 kN. The specimens were rigidly connected to the testing machine using bolts, and the test setup is illustrated in Fig. 7.2. White grids were marked on the surfaces of the columns to facilitate the observation of local buckling modes and locations of crack initiation, as shown in Fig. 7.3. Two laser displacement meters with a maximum displacement capacity of ±100 mm were placed between the two loading plates to measure their

Table 7.2 Mechanical properties and chemical composition of coupons

Coupons	Yield stress (N/mm^2)	Tensile strength (N/mm^2)	Uniform elongation (%)	Chemical composition (wt%) (from mill test certificate)				
				C	Si	Mn	P	S
RH1P	258.1	354.9	34.5	0.07	0.01	0.56	0.014	0.005
RH1C	271.0	354.4	35.5					
RH1W	242.0	332.2	33.9[a]					
RH2P	233.2	378.5	37.2	0.06	0.01	0.55	0.015	0.003
RH2C	244.6	383.6	40.6					
RH2W	230.0	372.0	26.9					
RH3P	238.3	340.9	46.8	0.07	0.01	0.56	0.017	0.003
RH3C	211.4	333.7	48.0					
RH3W	205.1	321.1	35.3					

Notes The last letter in the names of the coupons denotes the locations of the coupons, where "P" denotes the "parallel part," "C" denotes the "corner part," and "W" denotes the "welding part"
[a]The gage length of all the coupons is 50 mm except for Coupon RH1W which has a gage length of 25 mm

relative displacement, and the measured displacement data was employed to control the tests. Two stainless thin plates were located between the bolts and the upper loading plate to reflect the laser. A camera was placed at each side of the column to record the time and locations of crack initiation and propagation. Pictures were taken every five to fifteen seconds and timed with the corresponding load–displacement data.

7.2.3 *Loading Histories*

Four different loading histories, i.e., monotonic tension, a single full cycle, incremental and constant-amplitude loading, as illustrated in Fig. 7.4, were designed to simulate the ULCF loading history during an earthquake event. The first two loading histories were designed to simulate near-field earthquakes, where there is a large displacement pulse in the response of the structure. The last two histories were intended to simulate far-field earthquakes.

7.2.4 *Coupon Tests of SHS Columns*

To obtain the model parameters related with the aforementioned plasticity models and the fracture model, tension coupons were manufactured from the different locations

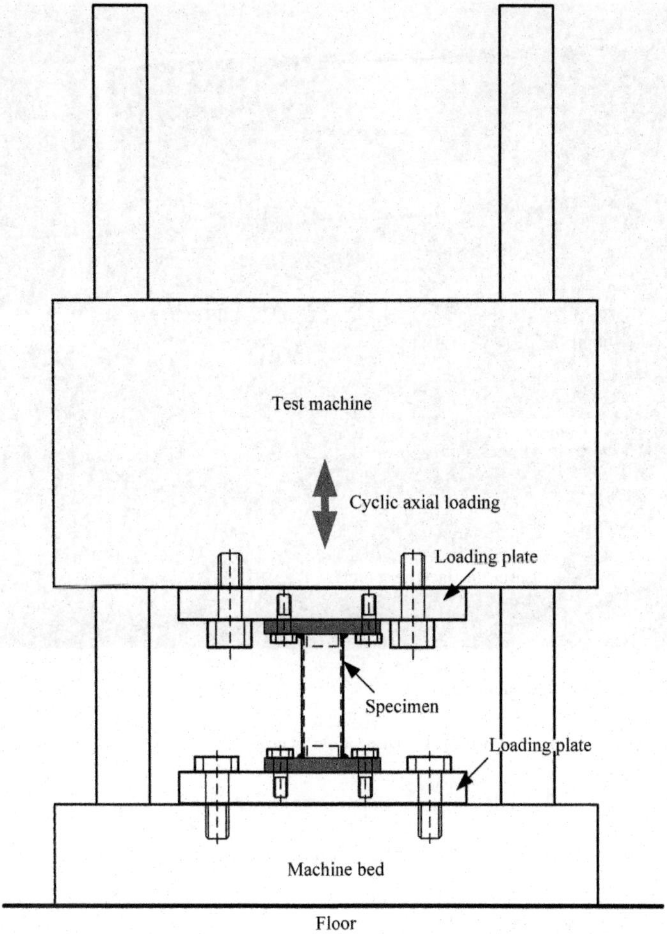

Fig. 7.2 Illustration of test setup

of the same SHS columns where the specimens were cut from. The material properties may be different from location to location over the whole section. Three kinds of coupons, i.e., coupons cut from the corner part, the parallel part, and the welding part as illustrated in Fig. 7.5, were manufactured to obtain the average material properties of the whole section. For the welding part, the inner weld beads of the original box columns are removed before manufacturing of the coupons. For each type of coupons, three coupons were made to study the material deviations and obtain the average properties of the materials. Totally 27 coupons are manufactured (Figs. 7.6 and 7.7).

The monotonic tension coupon tests were carried out with the setup as shown in Fig. 7.8. Two strain gauges were employed along the vertical direction of the two surfaces of the coupons to measure the strain data before necking occurs. An

Fig. 7.3 Measurements of specimens

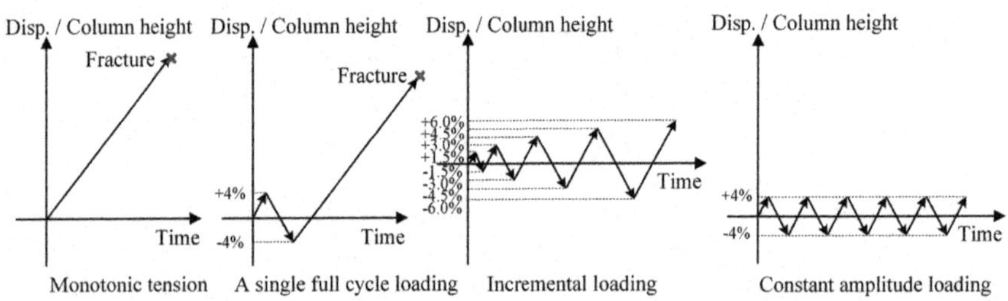

Fig. 7.4 Loading protocols of box columns

extensometer with a gage length of 50 mm was also employed in the tests to measure the extension and obtain the strain data after necking occurs, since the capacities of the gauges are around 20% which is close to the strain at necking initiation. The gage length is long enough to ensure the necking occurs within the length, and the extensometer was attached to the coupons tightly using the knife-edge clip-on devices. It was found that no slip occurred between the extensometer and the coupons for all the coupons even till the fracture. However, the measuring range of the extensometer is limited, and occasionally the maximum extension of the coupon may exceed the measuring capacity. The data obtained by the extensometer before necking occurs compares well with the data of the strain gauges.

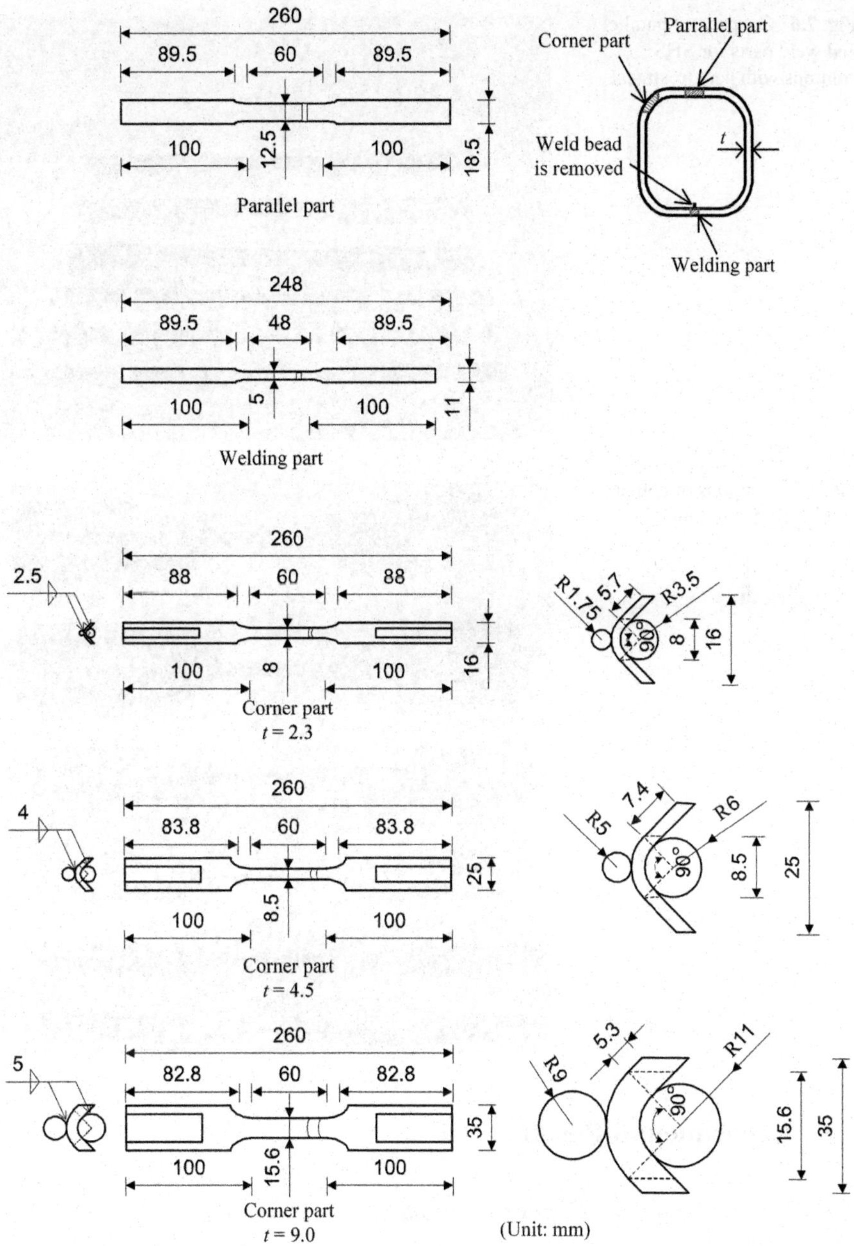

Fig. 7.5 Configurations of coupons for SHS columns

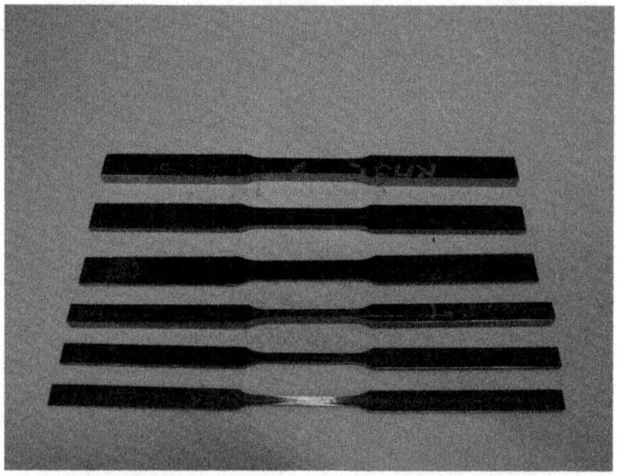

Fig. 7.6 Coupons of parallel and weld parts for SHS columns with heat treatment

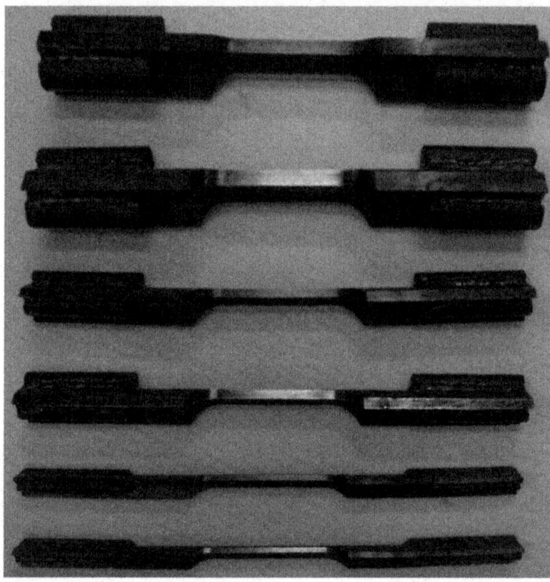

Fig. 7.7 Coupons of corner part for SHS columns

7.3 Experimental Results

7.3.1 Buckling and Cracking Modes

Buckling mode as illustrated in Fig. 7.9a was observed during the experiments. Crack initiation is determined by definition of a crack. Despite the fact that a micro-crack is defined in the scale of 0.01–0.1 mm, a crack in damage mechanics is commonly

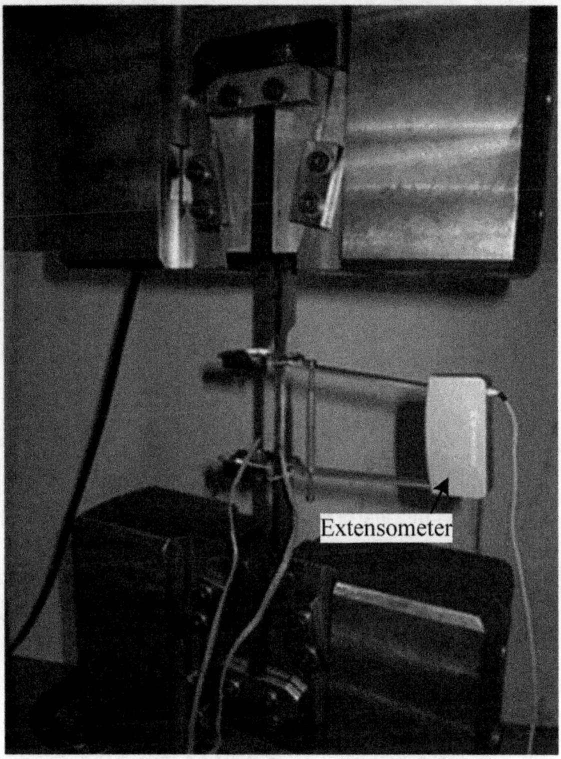

Fig. 7.8 Setup of monotonic tension coupon test of SHS columns

defined in the scale of around 1 mm (Chaboche 1988). Since a crack of 1 mm can be observed by human eyes as well as a digital camera, the damage mechanics definition is employed in this chapter.

Two categories of representative cracking modes, that are single-crack mode and multiple-crack mode, were identified in the experiments. The single-crack mode is characterized by a single macro-crack running through the whole section as illustrated in Fig. 7.10b. The multiple-crack mode, as illustrated in Fig. 7.9b, is characterized by several discrete cracks at the corners of the outer or inner surfaces.

The single-crack mode only occurs in specimens under monotonic tension and a single full-cycle loadings. There are some differences in the failure processes of the specimens with different width-to-thickness ratios. For specimens with a relative large width-to-thickness ratio of 47.6, thinning of the tube walls first occurs after the peak load is reached, followed by formation of an X-shaped shear band due to the plane stress state under increased tensile deformation, and a V-shaped crack finally forms along the shear band as the tensile displacement increases. The phenomenon of shear band formation has also been observed in very high strength CHS members by Jiao and Zhao (2001). However, for specimens with medium and small width-to-thickness ratios, there is no formation of X-shaped shear band due to the relatively

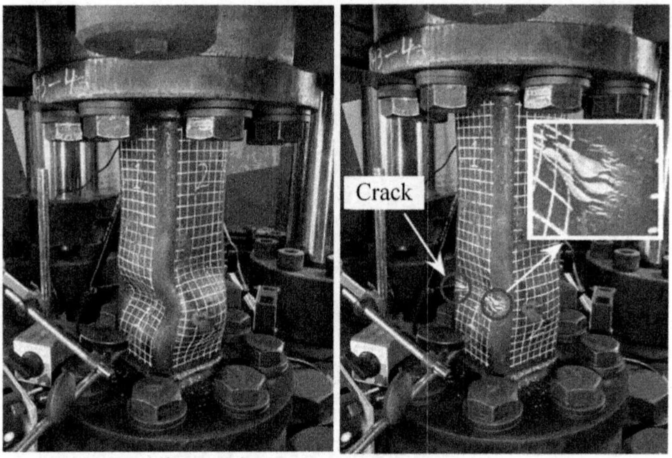

(a) Buckling mode (b) Multiple-crack mode

Fig. 7.9 Buckling and cracking modes of a heat-treated SHS stub column (specimen *RH3-4*)

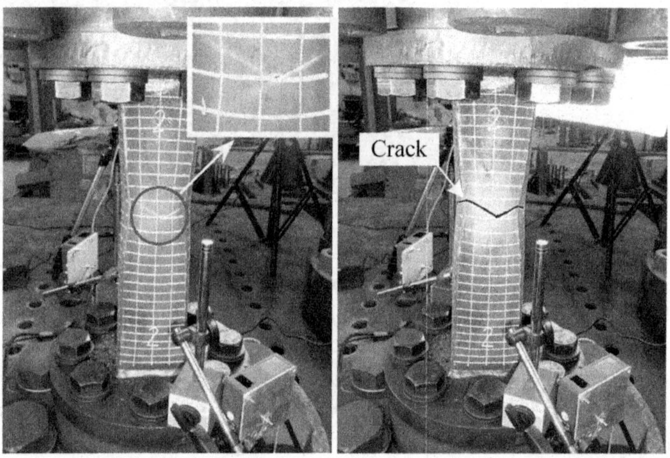

(a) X-shaped shear band (b) Single crack mode

Fig. 7.10 Formation of shear band and single-crack mode of a heat-treated SHS stub column (specimen *RH1-1*)

strong constraint in the thickness direction as the wall thickness increases. The crack along the width of the cross section is almost horizontal but not V-shaped.

The multiple-crack mode commonly occurs in specimens under incremental and constant-amplitude loadings. For this cracking mode, local buckling with the mode similar to Fig. 7.9a first occurs, where strain concentration occurs at the corners. Subsequently, several minor discrete cracks form at the corners of the specimens after several cyclic loading cycles. Several macro-cracks then initiate at the corners

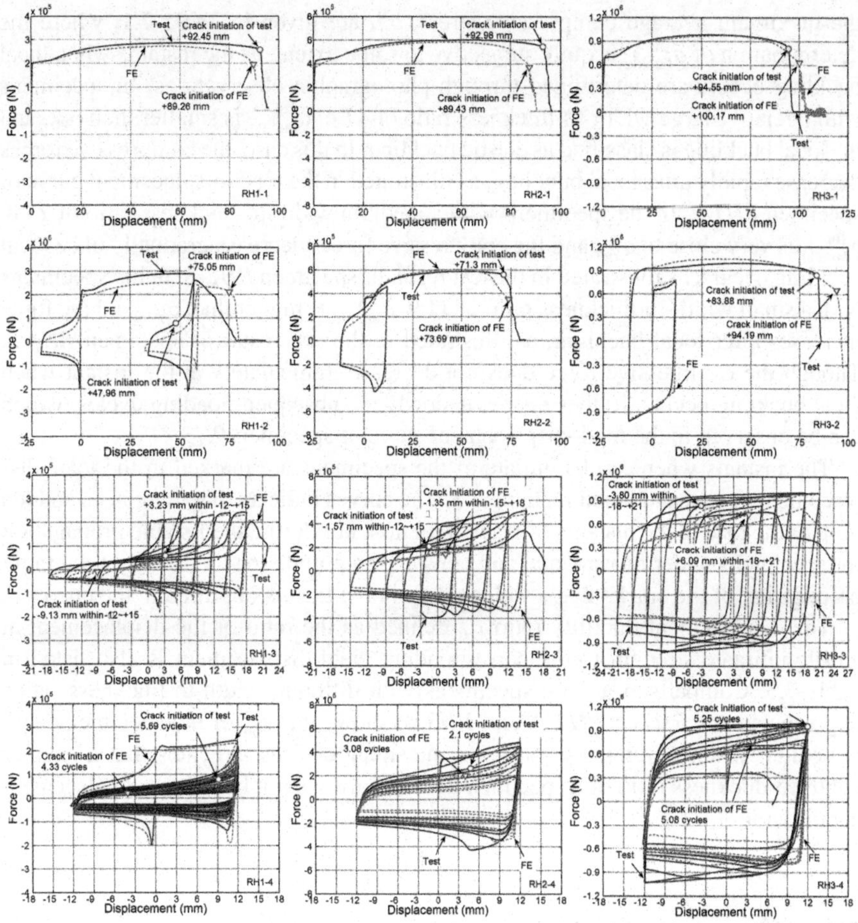

Fig. 7.11 Comparison of load–displacement curves of experimental and numerical results

of the outer surfaces of the concave walls, and also at the corners of the inner surfaces of the convex walls. Eventually, the multiple cracks form a major crack, and the experiments are ended when the major crack propagates quickly through the whole cross section.

7.3.2 Hysteretic Response

The load–displacement curves of the heat-treated specimens are shown in Fig. 7.11. Since the slenderness ratios of the specimens are very small, local buckling occurred in all the specimens except for the specimens under monotonic tension. The average initial tensile yield stress, σ_{y0}, the average initial local buckling stress, σ_{buckle}^{ini}, and the maximum average compressive stress, σ_u^c, are given in Table 7.1, where the determination of σ_{buckle}^{ini} is quite subjective to some extent, as the instants when local buckling occurs are determined through photographic observations. In specimens with a relative large width-to-thickness ratio of 47.6, σ_{buckle}^{ini} is smaller than σ_{y0}, and the local buckling is classified as elastic buckling. In this case, the compressive forces decrease rapidly after local buckling, as illustrated in the load–displacement curve of specimen *RH1-2*. In the specimens with a medium width-to-thickness ratio of 23.8, σ_{buckle}^{ini} is very close to σ_{y0}, and the compressive forces decrease gradually once local buckling occurs, as illustrated in the test result of specimen *RH2-2*. In the specimens with a small width-to-thickness ratio of 11.9, σ_{buckle}^{ini} is higher than σ_{y0}, and the force increases after local buckling, as illustrated in the test result of specimen *RH3-3*. Though the compressive force does not decrease immediately at the instant when local buckling occurs, it does decrease slowly in subsequent loading cycles, which can be observed in the load–displacement curve of specimen *RH3-4*.

The instants when cracks initiate of the specimens are marked in the load–displacement curves and listed in Table 7.3. The curves indicate that there is no sudden drop of the load at the instants of crack initiation, nor during their stable propagation for specimens under incremental and constant-amplitude loadings. It is at the rapid propagation of the cracks that a sudden drop in the load is observed in the curves.

The displacement ductility ratio μ_d defined as the ratio of the displacement at tensile ultimate strength to the corresponding yield displacement is also listed in Table 7.3. Comparison among specimens with different width-to-thickness ratios (e.g., specimens *RH1-3*, *RH2-3*, and *RH3-3*) shows that μ_d of the specimens with a medium width-to-thickness ratio are the smallest. This indicates that a smaller width-to-thickness ratio of a specimen does not always imply a larger displacement ductility ratio.

7.4 Numerical Simulation

One-fourth three-dimensional solid FE models as illustrated in Fig. 7.12 were established using ABAQUS (2010) based on the consideration of geometrical symmetry and reduction of computing time. The ABAQUS/Explicit module was selected to carry out the simulations, since the module can effectively handle severely nonlinear problems such as fracture, contact, and dynamic collision. The step times of the analyses were set long enough to ensure that the analyses be quasi-static and the inertial energy be neglectable compared with the total internal energy. Eight-node linear solid elements with reduced integration, *C3D8R*, were adopted in all the simulations.

The symmetric boundary conditions as shown in the figure were applied to the two symmetric surfaces. To simulate the constraints of the loading plates to the SHS columns, all the freedoms of the top and bottom surfaces were coupled to the centers (reference nodes in the figure) of the corresponding sections. Boundary conditions were applied directly to the reference nodes, where the bottom reference node is fixed, while all the freedoms except for the axial translational freedom were fixed

Table 7.3 Test results of specimens

Specimens	d_{crackini} (mm)	Loading cycle up to crack initiation	d_y (mm)	d_u (mm)	μ_d (mm)	${}_{h}E_p$ (kJ)	${}_{h}E_p/{}_{h}E_e$ ${}_{c}E_p/{}_{c}E_e$	Cracking modes
RH1-1	+92.45	–	0.38	92.45	243.29	24.5	645	Single-crack
RH1-2	+47.96	–		58.51	153.97	22.1	582	Multiple-crack
RH1-3	+3.23	−12 to +15 mm		19.11	50.29	17.8	468	Multiple-crack
RH1-4	+9.20	5.69 cycles		12.00	31.58	34.2	900	Multiple-crack
RH2-1	+92.98	–	0.34	92.98	273.47	52.1	840	Single-crack
RH2-2	71.30	–		71.30	209.71	52.4	845	Single-crack
RH2-3	−1.57	−12 to +15 mm		15.00	44.12	47.7	769	Multiple-crack
RH2-4	+4.80	2.1 cycles		12.00	35.29	52.8	852	Multiple-crack
RH3-1	+94.55	–	0.35	94.55	270.14	85.7	726	Single-crack
RH3-2	+83.88	–		83.88	239.66	116.7	989	Single-crack
RH3-3	−3.80	−18 to +21 mm		21.00	60.00	263.9	2236	Multiple-crack
RH3-4	+12.00	5.25 cycles		12.00	34.29	271.2	2298	Multiple-crack

Notes d_{crackini} denotes the displacement when cracks initiate; d_y and d_u denote the displacements corresponding to the yield force and tensile ultimate strength, respectively; μ_d is the ratio of d_u to d_y, ${}_{h}E_p$ is the total dissipated plastic energy till loss of tensile strength of the heat-treated specimens, ${}_{h}E_e$ is the corresponding energy till initial tensile yielding of each specimen

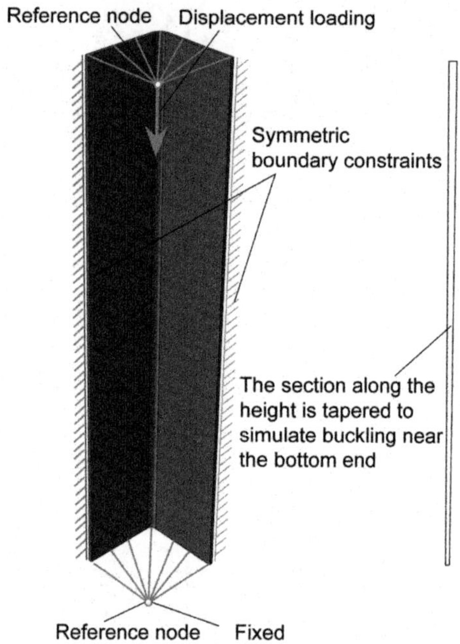

Fig. 7.12 One-fourth solid FE model of specimens

for the top reference node. Displacement loading histories same as those of the experiments were applied to the top reference node. For specimens under monotonic tension, perfect models were employed in the simulations. The test results show that local buckling does not always occur at the mid-height of the specimens. To simulate the asymmetric buckling mode, tapered cross section along the height direction as shown in Fig. 7.12 was employed to generate the local buckling at the bottom end. Meanwhile, buckling analyses of the model shown in Fig. 7.12 were carried out, and geometric imperfection with the distribution the same as the first-order eigenvector was introduced to the corresponding FE models to simulate the local buckling, where the maximum size of the imperfection was set to one-tenth of the corresponding wall thickness. The tube wall thickness was continuously reduced, which will have minor effects on the stress states. Ductile crack initiation was simulated using the cyclic fracture model and the plasticity model given in the previous chapters, and a simple element deletion method was employed to simulate crack propagation, where the same rule as that of the crack initiation was adopted. Since the self-weight of the specimens is relatively small compared with the ultimate strength, it was not considered in the simulation.

Numerical models with various element sizes as illustrated in Fig. 7.13 were established to study the effects of element size on the simulation results. The total numbers of elements of the models with coarse, medium, and fine mesh are 6840, 11,520, and 21,600, respectively. The load–displacement curves obtained using the three FE models for the specimen *RH2-3* are compared in Fig. 7.14. The three models generally give similar load–displacement curves, while the model with coarse mesh

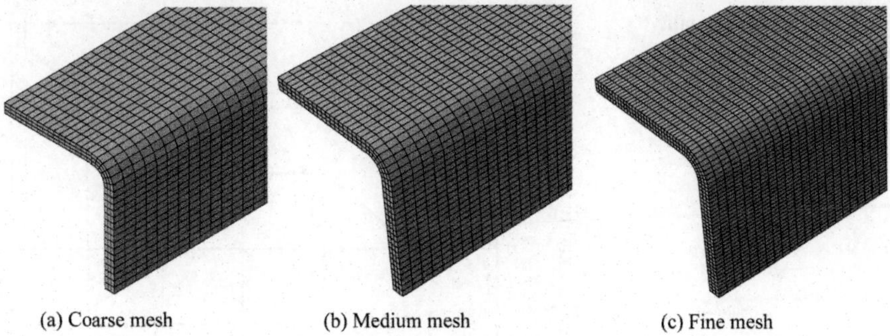

Fig. 7.13 FE mesh of a specimen (specimen *RH2-3*)

Fig. 7.14 Convergence study for FE models of heat-treated specimens (specimen *RH2-3*)

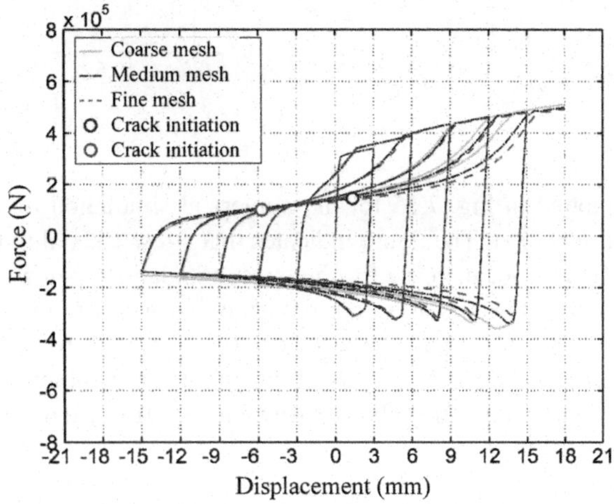

cannot simulate the buckling mode of the specimen accurately in reversal loading cycles, and therefore cannot catch an accurate location of crack initiation, which is therefore not marked in the curve. The predicted instants at crack initiation using the FE models with medium and fine mesh are of the same loading cycle, and the model with the finest mesh gives an earlier prediction of the crack initiation compared with the one with medium mesh. The numerical simulation for model with fine mesh takes a large amount of calculation time up to one week, and the medium-meshed model takes two to three days to carry out a simulation of the column under ULCF loading. It was found that the instants at crack initiation could be well evaluated when the mesh was fine enough to simulate the accurate buckling modes of the experiments. Therefore, the medium mesh was employed in the simulations of the heat-treated specimens under the consideration of both accuracy and efficiency.

The ductile fracture model for ULCF loading proposed in Chap. 6 was employed. To facilitate the understanding of the proposed cyclic fracture model, the stress triaxiality and the damage index versus equivalent plastic strain curves were respectively

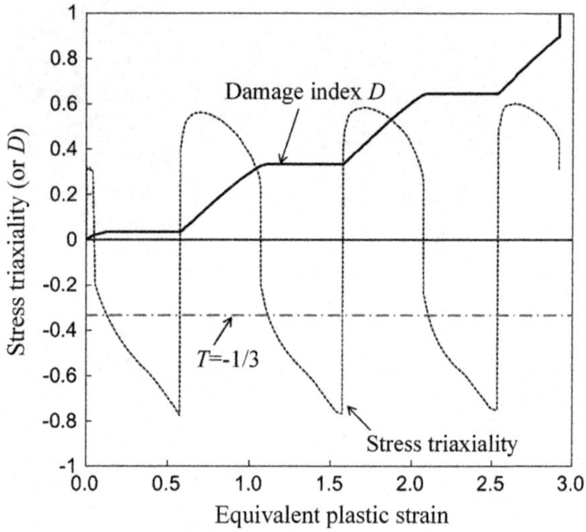

Fig. 7.15 Stress triaxiality and damage index versus equivalent plastic strain curves for *RH2-4*

plotted in Fig. 7.15 for the numerical simulation result of specimen *RH2-4* as an illustration. The figure indicates that stress triaxiality ranges from −0.8 to 0.6, and *D* does not increase for the stress triaxialities below −1/3. It is also interesting to find that *D* increases approximately linearly with the equivalent plastic strain.

The model parameter χ_{cr} can be calibrated from the monotonic tension coupon tests of the parrallel parts of the heat-treated columns using the approach proposed in a previous study (Jia and Kuwamura 2014a), and the procedure to obtain the value is as follows

(1) Obtain the true stress–true strain data up to fracture of the material using the weighted average method given in Chap. 2 from test results of a tension coupon;
(2) Carry out ductile fracture simulations of the coupon test using the true stress–true strain data in Step 1 and the ductile fracture model and find out the optimal value of χ_{cr} which gives the best evaluation of the instant at loss of tensile strength for the tension coupon test.

Tension coupons cut from the welding parts, parallel parts, and corner parts of the SHS tubes after heat treatment were carried out to obtain the values of χ_{cr} for the specimens. The material properties of the heat-treated specimens with the same thickness were found almost the same at the whole cross section, and the coupon test results of the parallel parts were employed to obtain values of the fracture parameter, which are listed in Table 7.4.

For the plasticity model, the Chaboche model with IH was employed, and the parameters were calibrated using the monotonic tension coupon tests according to the method proposed in Chap. 3 (Jia and Kuwamura 2014b). The values of the model parameters are listed in Table 7.4.

7.5 Comparison of Experiment and Simulation

7.5.1 *Comparison Results of Specimens with a Large Width-to-Thickness Ratio*

Whether the numerical simulation can catch the main characteristics of the failure processes of the specimens is of great importance. For the specimens with a large width-to-thickness ratio under monotonic loading, the experimental result shows that the failure process can be divided into the following steps as illustrated in Fig. 7.16: (1) yielding of the whole section; (2) reaching the ultimate tensile strength and initiation of thinning of the column walls near the mid-height of the specimen; (3) formation of shear band; (4) crack initiation; (5) rapid propagation of the crack and sudden loss of tensile strength. The crack propagation in the specimen under monotonic tension is so fast in the experiment that the photograph at crack initiation cannot be obtained precisely. The numerical results can well simulate the formation of the shear band and the configuration of the crack as shown in Fig. 7.17.

Table 7.4 Model parameters of Chaboche model and fracture model for specimens

Material	σ_{y0}	Chaboche model with isotropic hardening								Fracture model
		C_1	C_2	C_3	γ_1	γ_2	γ_3	k	Q_∞	χ_{cr}
RH1	258.1	46.0	50.9	643.1	0.0	0.0	8.0	1.5	298.7	3.2
RH2	233.2	100.5	582.6	562.9	0.0	11.0	11.3	1.3	341.2	3.5
RH3	238.3	59.7	26.2	657.3	0.0	0.0	7.6	0.8	365.0	3.5

Notes Unit of σ_{y0}, C_1, C_2, C_3, Q_∞: MPa
Unit of γ_1, γ_2, γ_3, k, and χ_{cr}: dimensionless

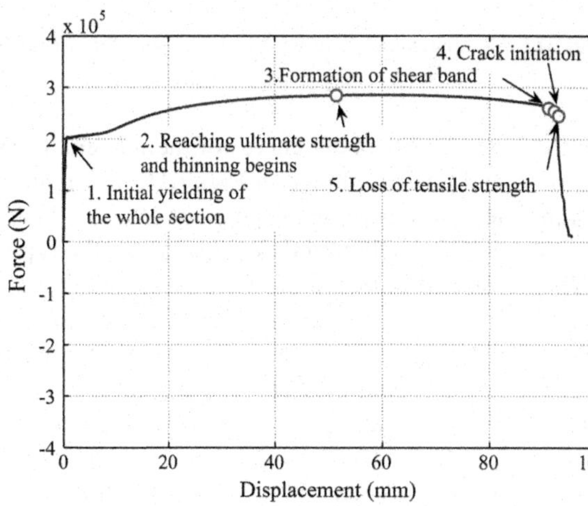

Fig. 7.16 Failure process of specimen *RH1-1*

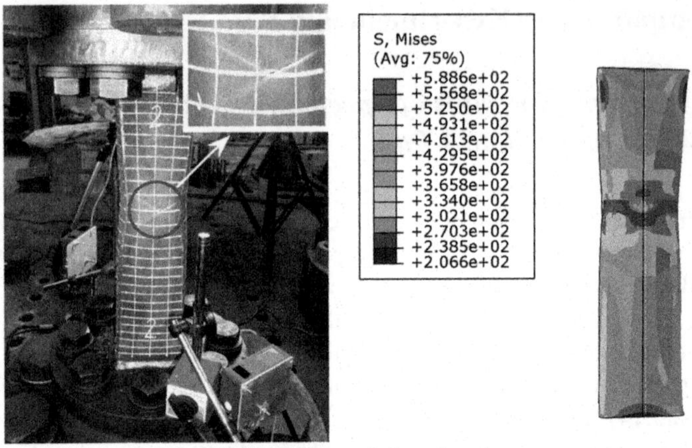

(a) Formation of shear band

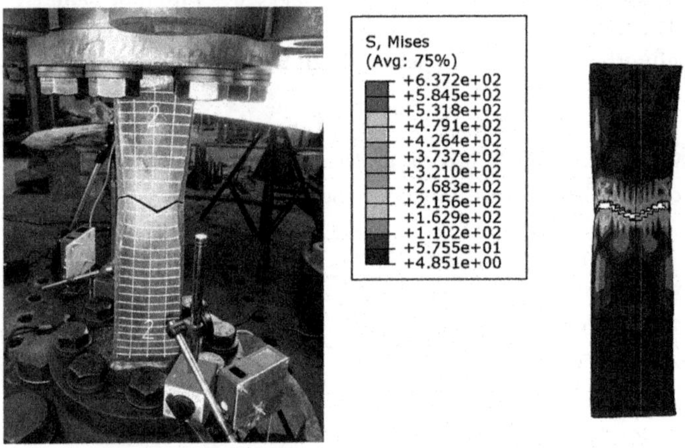

(b) Final configuration

Fig. 7.17 Comparison of failure process for specimen *RH1-1* (*Unit* MPa)

The load–displacement curves of the numerical results also agree well with the experimental ones as shown in Fig. 7.11. The numerical simulation also can well capture the failure process of specimen *RH1-2* which is similar to that of specimen *RH1-1*. The FE result slightly overestimates the rupture displacement of specimen *RH1-2* as shown in Fig. 7.11. This overestimation may be due to the fact that the cyclic fracture model assumes no damage when the stress triaxiality is below $-1/3$, which may underestimate the damage at extremely large plastic strain ranges under negative stress triaxialities. This underestimation becomes more apparent when the width-to-thickness ratio of the member is large, where a severe strain concentration will occur at the buckled regions after local buckling initiates.

For heat-treated specimens with a large width-to-thickness ratio under incremental and constant-amplitude loadings, the experimental results show that the failure

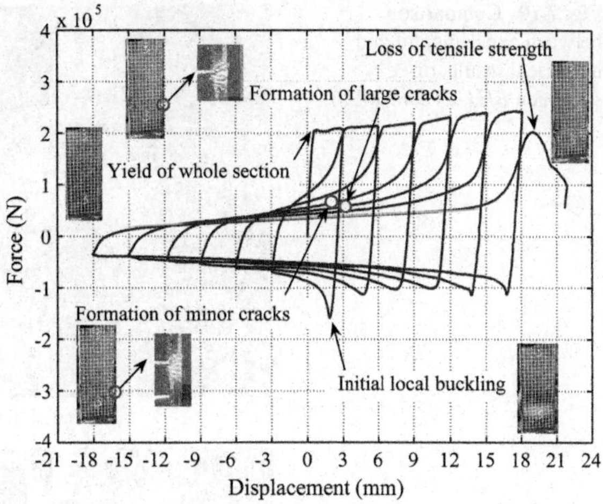

Fig. 7.18 Failure process of specimen *RH1-3*

process can be divided into the six steps as illustrated in Fig. 7.18: (1) yielding of the whole section under tension; (2) initial local buckling under compression; (3) formation of a series of minor cracks around the regions after being subjected to several large plastic loading cycles; (4) formation of macro-cracks under cyclic loading; (5) stable propagation of cracks under cyclic tension and compression; (6) rapid propagation of the cracks and sudden loss of tensile strength. The FE simulation can well predict the buckling mode and crack initiation location of the corresponding experiments as illustrated by the results of specimen *RH1-3* in Fig. 7.19. The simulations also can well evaluate the load–displacement curves and the instants at crack initiations as shown in Fig. 7.11, while it is difficult to accurately capture the instants when rapid propagation of the cracks occurs. The simulation of the cracked specimens under ULCF loading is much more complicated due to the problem of self-contact between the two cracked parts at the locally buckled regions under compression. Besides, the employed element deletion method cannot precisely capture the configurations of sharp crack tips, which may have some limitations for the simulation of crack propagation. The validity of the method should be further studied by both numerical and experimental studies. More accurate methods such as XFEM method (Belytschko et al. 2009) which can well describe the crack tip configurations may be required in future study. However, up to now there are still some limitations such as convergence problems for the XFEM method to apply to large plastic straining problems without pre-cracks, and the method is generally limited to elastic problems or cracked objects in commercial software (ABAQUS 2010).

Fig. 7.19 Comparison between experimental and numerical results of specimen *RH1-3* (*Unit* MPa)

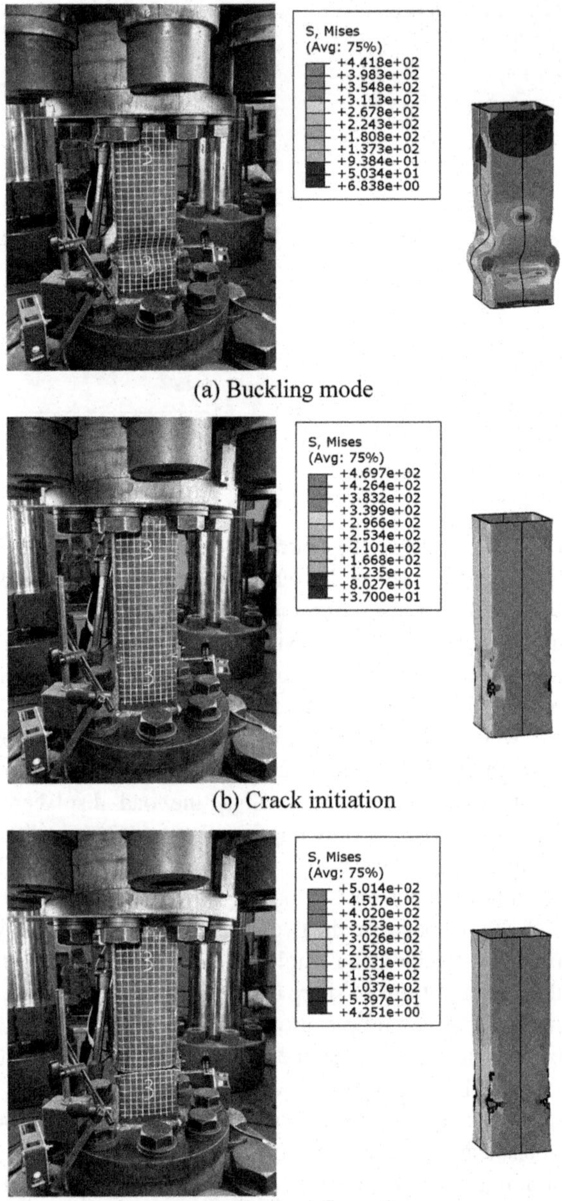

(a) Buckling mode

(b) Crack initiation

(c) Final configuration

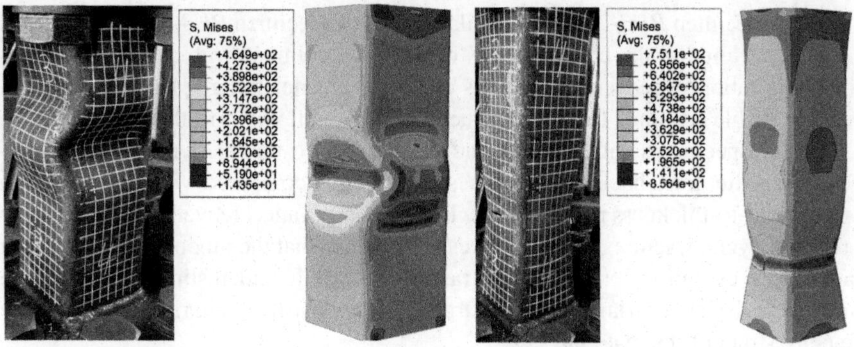

(a) Buckling mode (b) Crack initiation

Fig. 7.20 Buckling mode and crack initiation of specimen *RH2-2* (*Unit* MPa)

7.5.2 Comparison Results of Specimens with a Medium Width-to-Thickness Ratio

For the specimen with a medium width-to-thickness ratio under monotonic loading, the experimental result shows that the failure process is similar to that of the one with a width-to-thickness ratio of 47.6, where the main difference is that there is no formation of shear band before crack initiation. The load–displacement curve of the numerical result agrees well with the experimental one as shown in Fig. 7.11. The numerical simulation also can well simulate the failure process of specimen *RH2-2*, where crack initiates at the inflection point of the buckled specimen along the height as shown in Fig. 7.20. The load–displacement curve of the FE result for specimen *RH2-2* compares well with the test results as shown in Fig. 7.11.

The failure processes of specimens *RH2-3* and *RH2-4* are similar to those of specimens *RH1-3* and *RH1-4*, where cracks first occur at the corners of the specimens and propagate gradually under cyclic loading. The load–displacement curves of the FE results for specimens *RH2-3* and *RH2-4* compare well with the test results as shown in Fig. 7.11, while the instants at final rupture are not accurately captured by the simulations.

7.5.3 Comparison Results of Specimens with a Small Width-to-Thickness Ratio

For the specimen with a small width-to-thickness ratio under monotonic tension, there is also no formation of shear band before crack initiation. The load–displacement curves of the numerical results agree well with the experimental one as shown in Fig. 7.8. Since the buckling deformation is very small in the first cycle, the test result of specimen *RH3-2* is quite similar to that of specimen *RH3-1*. The numerical results well predict the crack instants of all the specimens with a small width-to-thickness ratio. Moreover, the instants at loss of tensile strength are also predicted with favorable accuracy for the specimens with a small width-to-thickness ratio.

For the specimens with medium and small width-to-thickness ratios, strain concentration after local buckling initiates is not so severe compared with the ones with a large width-to-thickness ratio. The crack initiation instants are well predicted by the proposed cyclic fracture model. This result indicates that the model can well predict the damage evolution at most of the strain ranges, while underestimates the damage under negative stress triaxialities at extremely large plastic strain ranges close to the fracture strain of the material.

7.6 Summaries

Experimental tests on twelve heat-treated square hollow section (SHS) stub columns were carried out to validate the applicability of the calibrated cyclic plasticity model and the cyclic ductile fracture model proposed in Chap. 6 to a complicated realistic case of post-buckling fracture of SHS columns. Two main factors, the width-to-thickness ratio and loading history were studied. Numerical simulations were also conducted to mainly simulate the ductile crack initiation of the specimens. The conclusions are drawn as follows.

(1) Two cracking modes, single-crack mode and multiple-crack mode, were observed in the experiments, where the single-crack mode occurred under monotonic tension and a single full cyclic loading, while the multiple-crack commonly occurred under incremental and constant-amplitude loadings.
(2) The cyclic plasticity model and the ductile fracture model can well predict the ductile crack initiation and the load–displacement relationships of the experiments. The two cracking modes were captured by numerical simulation, while the accurate shape of the multiple cracks was difficult to simulate due to the limitation of the element deletion method employed in the simulation.
(3) For the multiple-crack mode, one of the multiple cracks led to the final rupture, which was captured by the simulation, while the instants at the final rupture could not be captured accurately for some specimens by the element deletion method. The accuracy of the method for the crack propagation simulation needs further improvement.
(4) The instants at crack initiation predicted by the numerical simulations of the specimens with a large width-to-thickness ratio under a single full-cycle loading were slightly overestimated due to that the cyclic fracture model underestimates the damage under negative stress triaxialities in extremely large plastic strain ranges.

(5) The failure processes of the specimens with a large width-to-thickness ratio of 47.6 under monotonic tension were different from the others. A shear band formed before crack initiation due to the plane stress state within the thin walls.

References

ABAQUS (2010) ABAQUS standard manual (version 6.10). Karlsson & Sorensen Inc., Hibbitt, Pawtucket

Belytschko T et al (2009) A review of extended/generalized finite element methods for material modeling. Modell Simul Mater Sci Eng 17:043001

Chaboche JL (1988) Continuum damage mechanics: Part I—General concepts. J Appl Mech 55:59–64

Jia L-J, Kuwamura H (2014a) Ductile fracture simulation of structural steels under monotonic tension. J Struct Eng (ASCE) 140:04013115

Jia L-J, Kuwamura H (2014b) Prediction of cyclic behaviors of mild steel at large plastic strain using coupon test results. J Struct Eng (ASCE) 140:04013056

Jia L-J, Kuwamura H (2015) Ductile fracture model for structural steel under cyclic large strain loading. J Constr Steel Res 106:110-121

Jiao H, Zhao X-L (2001) Material ductility of very high strength (VHS) circular steel tubes in tension. Thin Wall Struct 39(887):906

Chapter 8
Post-buckling Ductile Fracture of Thin-Walled Steel Beam-to-Column Connections

Abstract Two thin-walled welded steel beam-to-column connections with weak panel zones are tested under ultra-low-cycle fatigue loading. Apparent shear buckling in panel zones is observed during the testing. The connections have stable and excellent seismic performance and finally fail due to post-buckling ductile fracture. This chapter aims to numerically simulate the post-buckling ductile cracking process using the previously proposed cyclic ductile fracture model and further investigate critical factors, such as equivalent panel zone width-to-thickness ratio, axial load ratio, and initial geometrical imperfection that affect cracking behaviors of the connections. The cyclic ductile fracture model is verified with the experimental data and employed to conduct the parametric analyses. Deterioration of stress-carrying capacity is also considered, and the fracture model can successfully simulate the load decrease in the hysteretic curves of the experimental results.

8.1 Introduction

A number of studies on design of panel zones in beam-to-column connections of welded steel moment-resisting frames (WSMRFs) have been conducted (Krawinkler 1978; Krawinkler et al. 1975; Popov et al. 1985). Panel zones are commonly designed strong enough to avoid premature buckling due to the shear load transferred by the adjacent beams, and this design concept has been accepted by most of the countries, such as the US and China (ANSI/AISC 341-10 2010; ANSI/AISC 360-10 2010; GB 50011-2010 2010). There are vertical connecting plates in the panel zones for spatial WSMRFs, which are used to link beam webs in the perpendicular direction. However, the favorable effect of vertical connecting plates on the shear stability of panel zones has not been considered in current design provisions. An experimental study has been conducted on the effect of vertical connecting plates in mitigating the shear buckling of panel zones in cruciform beam-to-column connections under cyclic loading (Pan et al. 2016). There are two types of connections in the experiments, where the connections with the vertical connecting plates are compared

L.-J. Jia and H. Ge, *Ultra-low-Cycle Fatigue Failure of Metal Structures under Strong Earthquakes*, Springer Tracts in Civil Engineering,
https://doi.org/10.1007/978-981-13-2661-5_8

with the ones without the vertical connecting plates. This chapter is focused on the latter, where the connections failed mainly due to post-buckling ductile fracture at base metal of the panel zones.

Ductile fracture is one of the most important failure modes in metal structures, where its fracture surface is featured with the dimple pattern. This failure mode is favorable to structural engineers, since a large amount of energy can be absorbed owing to plastic straining before rupture of the members or structures. Metallic dampers such as shear panels and buckling-restrained braces also mainly fail due to ductile fracture, and these structural components can thus absorb a large amount of energy transferred by the ground during a strong earthquake. Up to now, seismic performance of these structural components is still mainly evaluated using experimental studies due to lack of convenient and accurate ductile fracture models, especially for the cases where cyclic large plastic straining is involved.

A number of ductile fracture models (Jia and Kuwamura 2014; Kanvinde and Deierlein 2006; Liao et al. 2015; Panontin and Sheppard 1995; Rousselier 1987) are proposed based on the void growth theories (Mcclintock 1968; Rice and Tracey 1969). A damage plasticity model termed the Gurson model (Gurson 1975) and the GTN model (Tvergaard 1981; Tvergaard and Needleman 1984) is also widely employed to simulate ductile cracking of steel members and structures (Jia et al. 2016a, b; Liao et al. 2015; Qian et al. 2005). However, these models are mainly developed for monotonic loading, and their applicability to cyclic loading is still open to discussion. Up to now, limited research (Jia et al. 2016b; Jia et al. 2014; Jia and Kuwamura 2015; Kanvinde and Deierlein 2007, 2008) was carried out on ductile fracture under ultra-low-cycle fatigue (ULCF) loading, where cyclic plasticity, deterioration of stress-carrying capacity and extremely high geometrical nonlinearity are all concerned.

A two-parameter fracture model for ULCF has been proposed (Jia et al. 2016b), which consists of a crack initiation rule and a crack propagation rule. The crack initiation rule is a semiempirical formula based on both the void growth concept and former researchers' experimental findings (Bao and Wierzbicki 2005), where a ductile metal is postulated not to fracture when the stress triaxiality is less than a cutoff value of $-1/3$ (Jia and Kuwamura 2015). The crack propagation rule employs an energy approach, where the required energy to open a unit area crack is assumed to be a material constant and can be determined using standard V-notched Charpy impact tests (Jia et al. 2016b). The fracture model works well for three structural steels under cyclic combined shear and normal stress loading. This chapter aims to further validate the validity of the model using two full-scale welded steel cruciform beam-to-column connections under ULCF.

In this chapter, the fracture model and the experimental results are first briefly introduced. Subsequently, numerical analyses of the specimens are carried out to calibrate the numerical models, the fracture model, and the plasticity model. Finally, the failure mechanism of the specimens is explained based on the numerical results, and further parametric studies are conducted to investigate the factors that have not been comprehensively studied in the experiments. The effects of equivalent panel zone width-to-thickness ratio, axial load ratio, and initial geometrical imperfection on seismic performance, crack initiation, and propagation of the connections are investigated.

8.2 Two-Parameter Ductile Fracture Model for Cyclic Large Strain Loading

The ductile fracture initiation rule for ULCF loading proposed in Chap. 6 was employed. The damage index for crack initiation, D_{ini}, is defined

$$\mathrm{d}D_{\mathrm{ini}} = \begin{cases} \dfrac{\mathrm{d}\varepsilon_{\mathrm{eq}}^{p}}{\chi_{\mathrm{cr}} \cdot e^{-\frac{3}{2}T}} & T \geq -1/3 \\ 0 & T < -1/3 \end{cases} \tag{8.1}$$

where χ_{cr} is a material constant, $\mathrm{d}\varepsilon_{\mathrm{eq}}^{p}$ is the incremental equivalent plastic strain, and T is the stress triaxiality. χ_{cr} can be obtained through numerical simulation of the corresponding tension coupon test according to the procedure is given in Chap. 4, where the post-necking modification of the true stress–true strain data is required. Ductile fracture initiation of a material is postulated to occur when D_{ini} reaches one unit. Herein, the term "ductile fracture initiation" denotes micro-crack initiation at a scale of 0.01–0.1 mm, which cannot be visually observed during an experiment. Meanwhile, macro-crack initiation is defined at the instant when the crack length is larger than 1 mm and is termed as "crack initiation" in this study.

A crack propagation rule based on an energy balance approach has also been proposed in a previous study (Jia et al. 2016b). The crack propagation index, D_{prop}, is defined in the following formula,

$$D_{\mathrm{prop}} = \frac{G}{G_c} \tag{8.2}$$

where G is the current absorbed energy of a unit area since the instant of fracture initiation, and G_c is the threshold value for absorbed energy to open a unit area crack. G_c is postulated to be a material constant, and a simple method to obtain it is proposed using a conventional coupon under monotonic tension at a quasi-static speed. The deterioration of stress-carrying capacity is also considered based on an effective stress concept, where the current effective stress, σ_e, is defined

$$\sigma_e = (1 - D_{\mathrm{prop}}) \cdot \sigma \tag{8.3}$$

The loading and unloading tangent modulus of a damaged material, E_d, is reduced to

$$E_d = (1 - D_{\text{prop}}) \cdot E \tag{8.4}$$

where E is the Young's modulus of undamaged materials.

8.3 Experimental Study on Welded Beam-to-Column Connections

8.3.1 *Configuration of Specimens*

The two specimens with the configuration shown in Fig. 8.1 are made of Chinese Grade Q345B steel, a low alloy structural steel with a nominal yield stress of 345 MPa. The columns and beams are all H-shaped sections welded of steel plates. The nominal and measured sizes are listed in Table 8.1. The beams were welded to the column using complete joint penetration (CJP) welds. Three coupons from each plate thickness were tested to obtain the material properties. The average measured material properties of each steel plate are given in Table 8.2.

In both the American and Chinese seismic design provisions of steel structures, the following requirements are proposed to avoid premature shear buckling of the panel zone:

$$\begin{cases} t_{\text{pz}} \geq (h_{\text{pz}} + b_{\text{pz}})/90 \\ h_{\text{pz}} = h_{\text{wb}},\ b_{\text{pz}} = h_{\text{wc}} \end{cases} \tag{8.5}$$

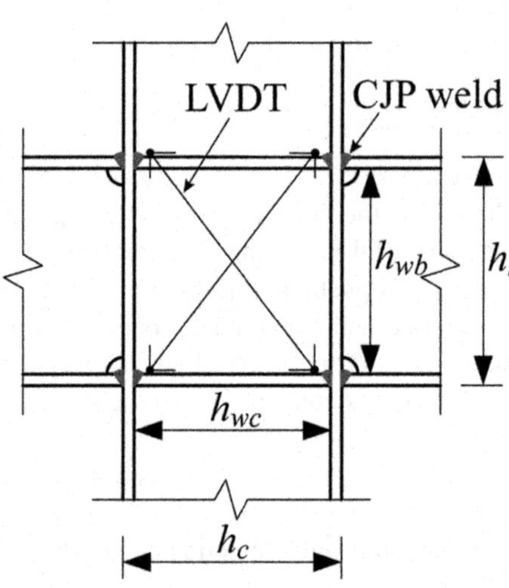

Fig. 8.1 Configuration of specimens at connections

Table 8.1 Geometrical and mechanical properties of specimens

Specimen		Column	Beam	Continuity plate	n	η
H1n3	Designed	H 270 × 175 × 6 × 8	H 270 × 125 × 5 × 6	254 × 60 × 6	0.3	–
	Measured	H 265 × 175 × 5.5 × 7.7	H 267 × 125 × 4.9 × 5.5	254 × 60 × 5.49	0.29	0.45
H2n3	Designed	H 300 × 175 × 5 × 8	H 300 × 125 × 5 × 6	284 × 60 × 6	0.3	–
	Measured	H 301 × 175 × 4.9 × 7.7	H 230 × 125 × 4.9 × 5.5	284 × 60 × 5.49	0.29	0.43

Notes n = axial load ratio, η = ratio of panel zone plastic shear strength to the beam plastic flexural strength

Table 8.2 Measured material properties

Nominal t (mm)	Measured t (mm)	f_y (N/mm^2)	f_u (N/mm^2)	Elongation (%)	Applicable locations
5	4.9	347	499	35.0	Beam webs; column web of specimen H2n3
6	5.5	358	494	33.1	Beam flanges; continuity plates; column web of specimen H1n3
8	7.7	386	534	29.8	Column flanges

where h_{pz} and b_{pz} are, respectively, the depth and width of the panel zone; h_{wb} and h_{wc} are depths of the beam and column webs. An equivalent panel zone width-to-thickness ratio, λ_{pz}, is defined according to Eq. (8.5) in this study.

$$\lambda_{pz}=(h_{pz}+b_{pz})/t_{pz} \tag{8.6}$$

In this study, λ_{pz} of the two specimens H1n3 and H2n3 are 92 and 118. Specimen H2n3 is susceptible to premature shear buckling.

Weak panel zones are designed for the two specimens to drive plastic deformation in the panel zones in this study. The ratios between panel zone plastic shear strength and the beam plastic flexural strength, η, of H1n3 and H2n3 are, respectively, 0.45 and 0.43. Therefore, the plastic deformation will mainly develop in the panel zones without apparent plastic straining at the beam ends. Moreover, in order to investigate the effect of the column axial load ratio, n, on the seismic performance of panel zones, the nominal n of the specimens are designed as 0.3, and the actual measured values are 0.29 due to the negative manufacturing tolerance of the plate thickness.

8.3.2 Test Setup and Loading Protocol

The test setup is illustrated in Fig. 8.2, where three loading devices were employed to conduct the quasi-static cyclic testing. Spherical bearings were employed at both column ends to simulate hinged boundary conditions at the mid-height of frame columns. The two beam ends were also pin-connected to the two vertical loading actuators with capacities of 100 kN/500 mm and 200 kN/500 mm, respectively. The column axial load was applied through a vertical loading jack with a capacity of 1000 kN at the top column end. Lateral supports were also employed at a distance of 500 mm away from the corresponding beam ends to prevent lateral buckling of the beams. A supporting brace was employed to connect the top spherical bearing with a reaction frame. The brace was rigidly connected to the spherical bearing, while hinged to the reaction frame. The out-of-plane supports were applied to the top spherical bearing to restrain the out-of-plane displacement of the top column end.

During the test, the axial compressive load was first applied to the top column end and kept constant at a designed stress ratio of 0.3 during the whole test. Subsequently, antisymmetric loading (−1:1) was applied to both beam ends simultaneously to generate high shear stress in the panel zone until its failure. Therefore, the panel zone was subjected to a combination of constant compression and incremental cyclic shear strain. To simulate the large plastic deformation history during a strong earthquake, the loading history shown in Fig. 8.3 is employed. Before yielding of the panel zone, the cyclic antisymmetric loading was controlled by the loads, and after yielding by the beam end displacement. In the displacement-controlled stage, each displacement level was cyclically loaded twice, and the displacement increment between two adjacent displacement levels was set as the yield displacement of the connections, which is close to 10 mm in this study. The experiments were all terminated when cracks ran through the plate thickness of the shear panels. The ultimate state is defined as the instant when the load-carrying capacities decrease by more than 15% compared with the corresponding peak loads.

8.3.3 Buckling and Fracture Modes

The failure modes of the two specimens as shown in Figs. 8.4 and 8.5 are similar to each other, i.e., ductile fracture of the panel zones due to repetitious shear buckling under the cyclic large strain loading. The crack initiation and propagation process of specimen H2n3 on the concave side are shown in Fig. 8.6. Tension and compression

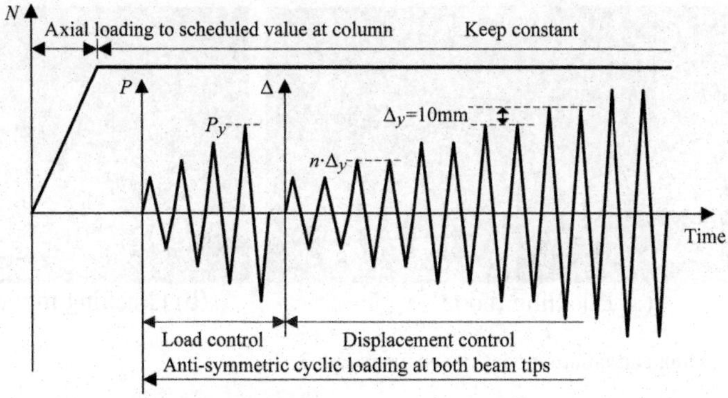

Fig. 8.3 Loading history for specimens

Fig. 8.4 Failure mode of specimen H1n3

diagonals, respectively, formed due to shear buckling of the panel zone. The angles of the diagonals to the horizontal direction were close to 45° since the width and height of the panel zones were designed the same. The ductile fracture is mainly due to flexural deformation of the panel zone induced by shear buckling. Cracks initiated at the concave side of the plate surface as illustrated in Fig. 8.7 and gradually ran through the whole plate thickness of the panel zone. In addition, just as pointed out by previous researchers, sharp change in curvatures of panel zone triggered local kinking at the column and beam flanges as shown in Fig. 8.4. This local kinking deformation may sometimes induce cracking at the welds of the beam-flange-to-column-flange joints. This is also one reason why the weak panel zone design was not recommended in past studies. Fibrous cracks were also observed at the weld toes as illustrated in Fig. 8.4, while the cracks did propagate apparently, but has no significant effect on the load–displacement curves of the connections.

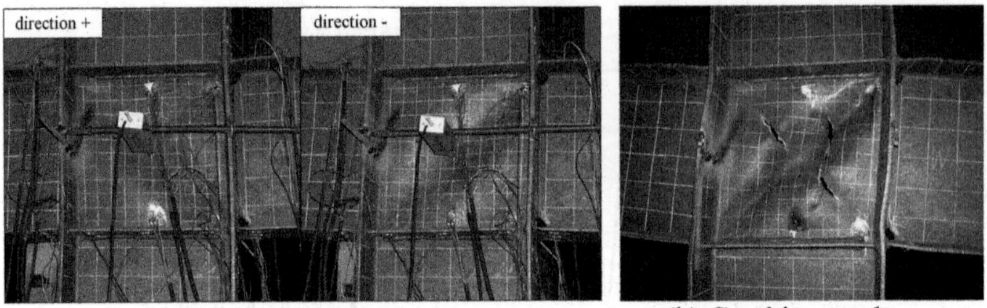

(a) Buckling mode (b) Cracking mode

Fig. 8.5 Buckling and failure modes of specimen H2n3

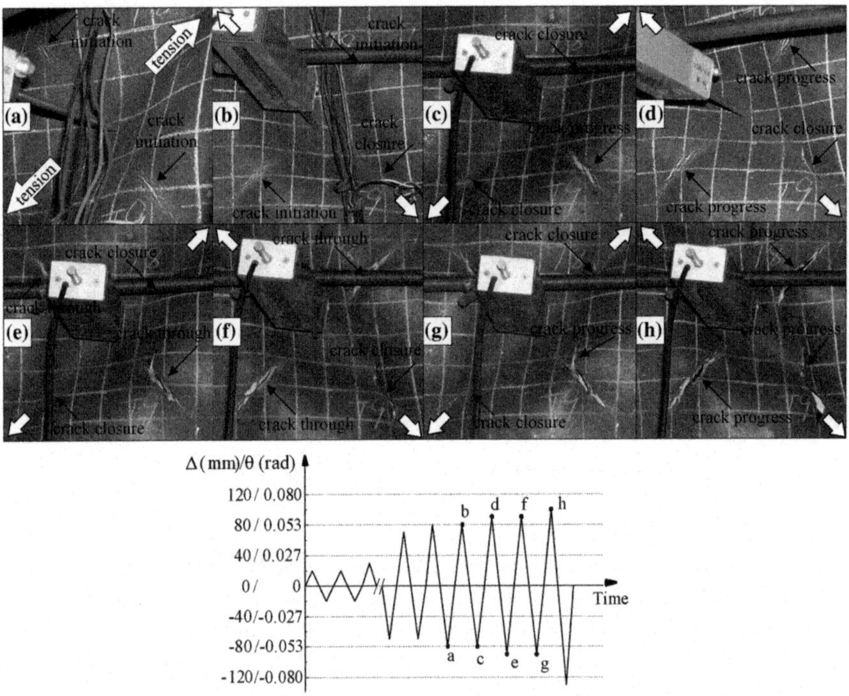

Fig. 8.6 Failure process of specimen H2n3 on concave side

8.3.4 Hysteretic Response

The hysteresis and corresponding skeleton curves of moment–rotation (M–θ) response and hysteresis curves of panel zone shear force–distortion (Q–γ) response of the two specimens are, respectively, given in Figs. 8.8 and 8.9, in which M is the beam moment at the column face, θ is calculated by the displacement difference of both beam ends over beam length, Q is panel zone shear force by unbalanced

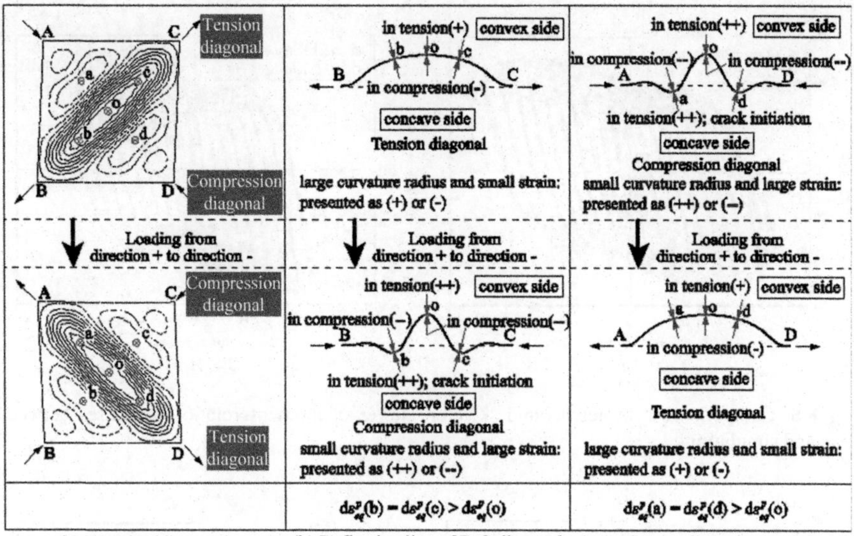

(a) Shear buckling mode (b) Deflection line of B-C diagonal (c) Deflection line of A-D diagonal

Fig. 8.7 Crack initiation point analysis

beam moments, and γ is obtained from the diagonal LVDTs as shown in Fig. 8.1. The hysteretic curves are stable and large rotation capacities far beyond 0.03 rad are achieved for both of the specimens, indicating favorable energy dissipation properties of the connections with weak panel zones. Besides, apparent pinching effect was also observed, which is commonly found in the load–displacement curves of a metallic shear panel damper. The instants of shear buckling and crack initiation of the panel zones are also marked in the skeleton curves of Fig. 8.8. This implies that both the shear buckling and the ductile crack initiation did not lead to any noticeable drop in the connection strength. Ductile crack propagation in the panel zones induced gradual decrease of the connection strength, but no sharp drop of the load was observed in the curves.

8.4 Numerical Simulation

8.4.1 FE Modeling

The experimental results were simulated using three-dimensional shell models as shown in Fig. 8.10 in the standard module (ABAQUS 2010). Shell elements S4R with reduced integration schemes were adopted in the static analyses for both accuracy and efficiency, and five integration points in the thickness direction were employed. The freedoms of all the nodes at column and beam ends were coupled to individual

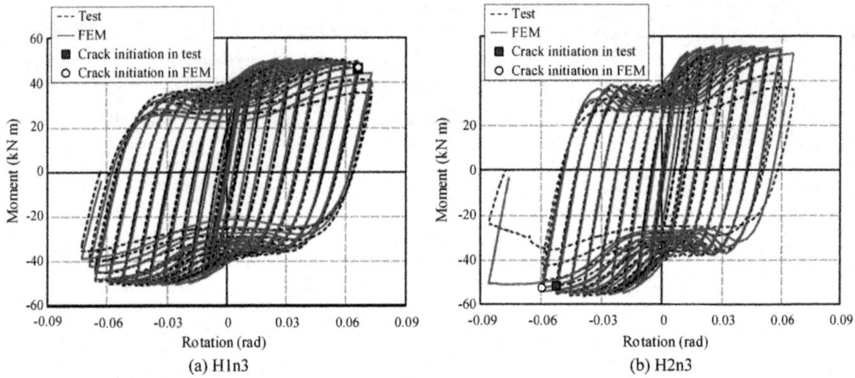

Fig. 8.8 Comparison of hysteresis and skeleton curves of moment–rotation responses between tests and simulations

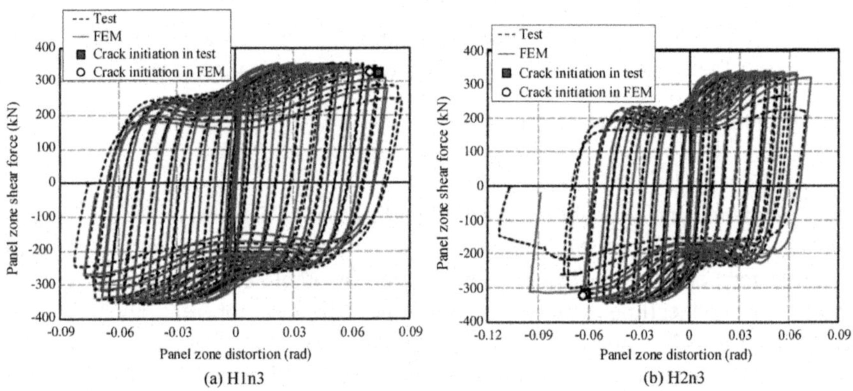

Fig. 8.9 Comparison of hysteresis panel zone shear force–distortion curves between tests and simulations

reference points, and boundary conditions close to the experiments were applied to the reference points. The sections where the lateral supports locate were also restrained. The procedure for the application of the loads is the same as that of the experiments.

Ductile cracking of the specimens was simulated by an element deletion approach in ABAQUS with the aforementioned two-parameter ductile fracture model, where an element is removed when the damage index, D_{prop}, reaches one. The Chaboche model with three backstresses was employed to simulate cyclic plasticity of the structural steel Q345B, and the model parameters are given in Table 8.3 based on the aforementioned tension coupons and previous cyclic coupon test results (Zhou et al. 2015). The fracture parameters are obtained based on the tension coupons, which is also given in Table 8.3. The welds were not incorporated in the model since no apparent failure was observed during the experiments. A convergence study on the

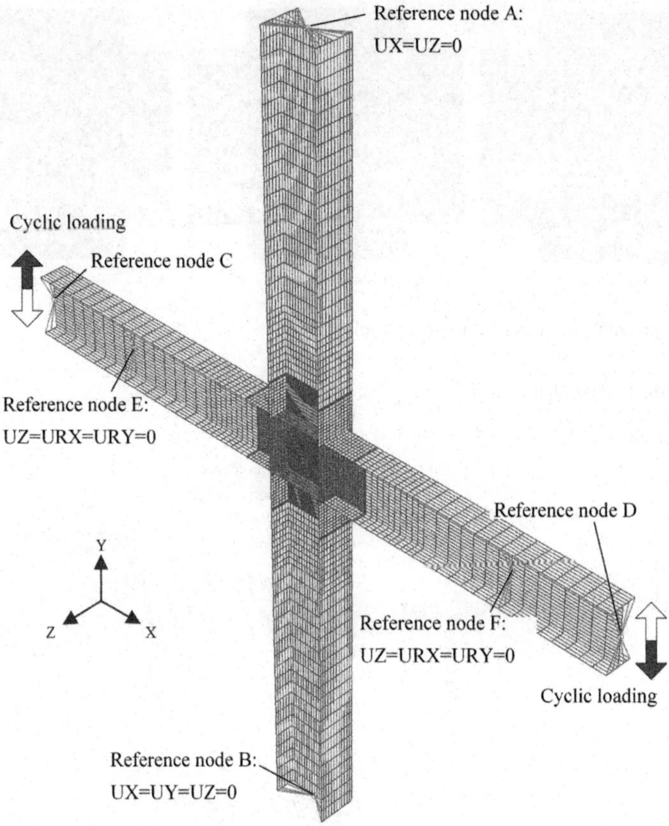

Fig. 8.10 FE model of connections

Table 8.3 Parameters for fracture model and Chaboche model in ABAQUS

E (MPa)	f_y (MPa)	C_1 (MPa)	γ_1	C_2 (MPa)	γ_2	C_3 (MPa)	γ_3	χ_{cr}	G_c (J/mm^2)
206,000	Measured value	31,864	522	4424	82	1030	0	2.2	0.155

effect of element sizes on the predicted moment–rotation curves was carried out, and three meshing schemes of the panel zone for specimen H1n3 as shown in Fig. 8.11 were employed. The comparison results of the obtained moment–rotation curves and required CPU time, etc., are given in Fig. 8.12 and Table 8.4. The medium mesh size was finally employed in the subsequent parametric analyses considering both accuracy and efficiency. Mesh size was found to have minor effect on the simulation results.

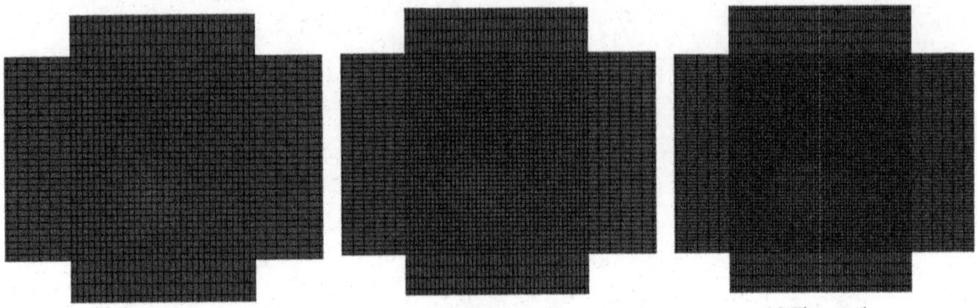

Fig. 8.11 Different mesh sizes for shear panel

Fig. 8.12 Mesh dependence effect of cracking simulation

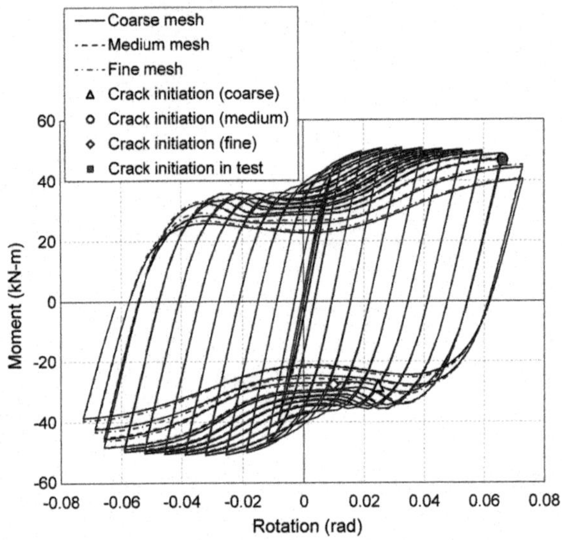

8.4.2 Comparison between FE and Experimental Results

The comparison between the experiments and the simulation results for the moment–rotation curves and the panel zone shear force–distortion curves of the two specimens are, respectively, shown in Figs. 8.8 and 8.9, where the curves compare well with each other. The predicted and actual crack initiation instants of specimen H1n3 match well, and the differences between the predicted and actual crack initiation instants of specimen H2n3 are within only a single displacement level. The predicted cracking modes are compared with the experimental ones in Fig. 8.13, where the cracking modes are quite similar to each other, and the comparison result of specimen H1n3 is better than that of specimen H2n3. The aforementioned results indicate good performance of the ductile fracture model under ULCF loading. The contour plot of the damage index, D_{ini}, of specimen H1n3 shown in Fig. 8.13a indicates that damage concentrates at both the diagonals and perimeter of the panel zone.

Table 8.4 Comparison of results from models with different mesh sizes

	Test	Coarse	Medium	Fine
Ultimate moment (kN m)	50.86	50.87	50.75	51.01
θ level at crack initiation (rad)	+0.066 (2nd cycle)	−0.066 (2nd cycle)	+0.066 (2nd cycle)	−0.066 (2nd cycle)
Accumulated energy from M–θ response (kJ)	96.82	88.76	88.51	–
Total elements	–	8908	10,884	14,704
Total nodes	–	9083	11,063	14,819
Total CPU time (min) [4CPU, 8G memory]	–	109	159	252
Wallclock time (min) [4CPU, 8G memory]	–	40	55	84

This also explains the fibrous cracks at the welds of the perimeter. The deformed configuration of the panel zones at the instant of crack initiation is given in Fig. 8.7 according to the analysis result. Cracks were found to initiate and propagate at the tension side of the point with the minimum curvature radius and thus the maximum strain amplitude as shown in Fig. 8.7. The contour plots of the damage index at the convex and concave sides before crack initiation are given in Fig. 8.14, implying that damage at the concave side is more remarkable than that at the convex one. The damage index versus the equivalent plastic strain curves and the stress triaxiality versus the equivalent plastic strain curves are, respectively, presented in Figs. 8.15 and 8.16. It can be found that D_{ini} increases almost linearly when the stress triaxiality is positive, and no damage accumulates when T is less than $-1/3$ according to the proposed ductile fracture model. It is also interesting to find that the peak positive and negative stress triaxialities of the two specimens are respectively close to a constant, which respectively equals to about 2/3 and $-2/3$. This explains why D_{ini} increases almost linearly for positive stress triaxialities. This finding is of great importance, since it indicates that the damage of the panel zone can be evaluated by the equivalent plastic strain at tensile loading half cycles with no need to consider the stress states. Figure 8.15 also shows that the equivalent plastic strain of H1n3 till fracture initiation is almost the same as that of H2n3 with a larger equivalent panel zone width-to-thickness ratio, indicating that the equivalent panel zone width-to-thickness ratio has a minor effect on the ductile fracture initiation.

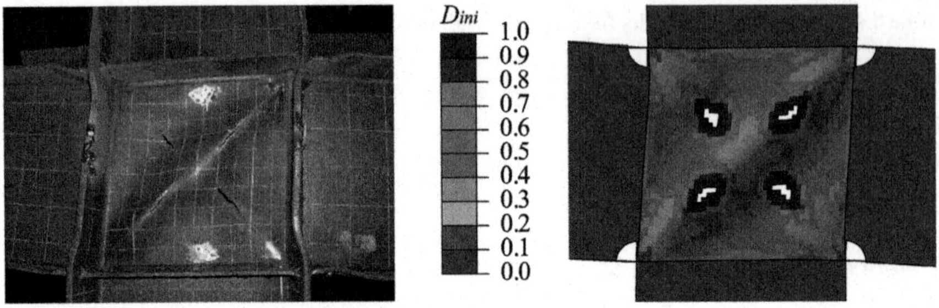

(a) North (convex) side of H1n3

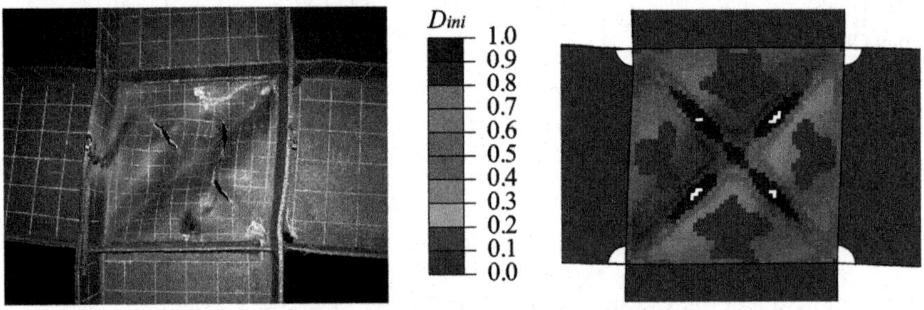

(b) North (concave) side of H2n3

Fig. 8.13 Comparison of fracture modes between tests and simulations

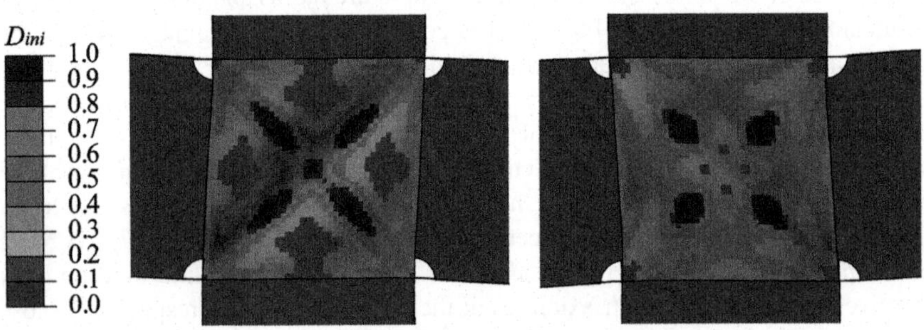

(a) South (concave) side (b) North (convex) side

Fig. 8.14 Contour plots of damage initiation index before cracking initiation of H1n3

8.4.3 Merits for Incorporating the Ductile Fracture Model

Numerical simulations without the implementation of the ductile fracture model were also conducted to evaluate the effect of cracking simulation on the moment–rotation curves of the two specimens. Comparison results of the numerical simulations with and without implementation of the ductile fracture model are given in Fig. 8.17. The figure indicates that deterioration of the load-carrying capacity can only be well evaluated when the ductile fracture model is incorporated. Moreover, the cracking

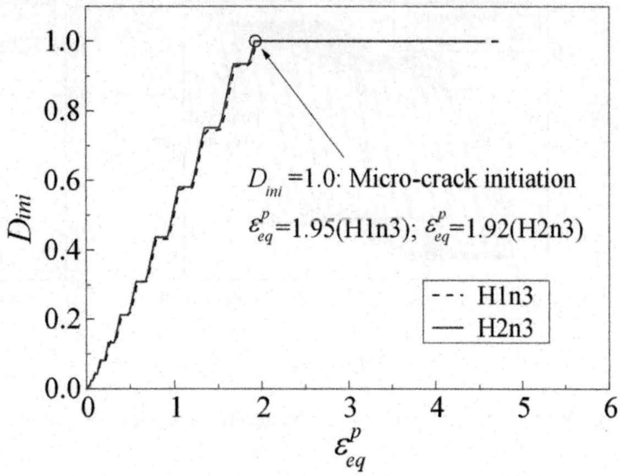

Fig. 8.15 Evolution of the damage index for two specimens

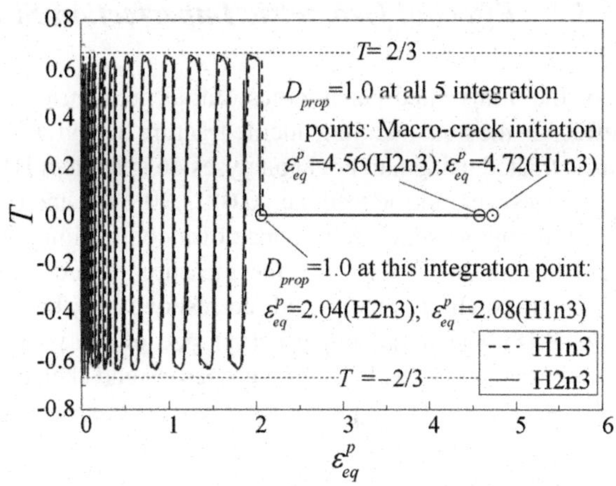

Fig. 8.16 Stress triaxiality histories for firstly removed elements

phenomenon observed during the experiments can also be well reproduced in the simulations with implementation of the ductile fracture model.

8.5 Parametric Analyses

In this section, parametric analyses using the calibrated shell models were conducted to investigate the effects of the geometric imperfection size, the axial load ratio, and the equivalent panel zone width-to-thickness ratio on the post-buckling fracture behaviors of the beam-to-column connections. The boundary and loading conditions are the same as those in Sect. 8.4 as illustrated in Fig. 8.10. The loading protocol is also the same as illustrated in Fig. 8.3. Likewise, the Chaboche model and the proposed ductile fracture model are employed with the same model parameters adopted.

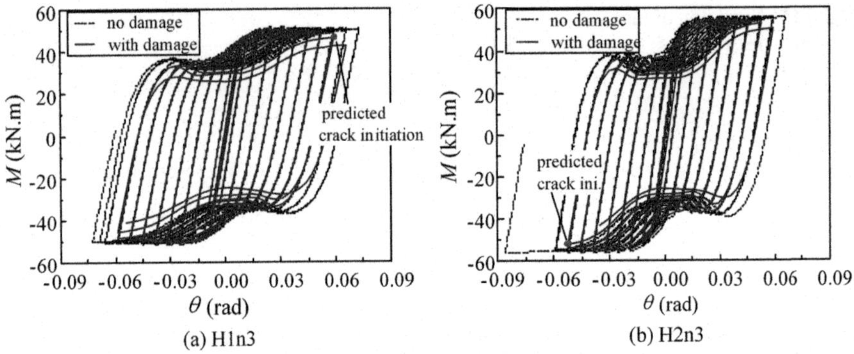

Fig. 8.17 Effect of cracking on hysteretic behaviors of specimens

8.5.1 Effect of Geometric Imperfection Size

For the thin-walled panel zones, initial geometric imperfection size may have an effect on the post-buckling ductile fracture behavior of the connections and was thus investigated using numerical analyses on specimen H1n3. Three different imperfection sizes were introduced, i.e., 0.01 t_{pz}, 0.1 t_{pz}, and 0.5 t_{pz}. The size of 0.01 t_{pz} was to study the seismic performance of a connection with minor initial imperfection, and the size of 0.5 t_{pz} was the measured value. The comparison results for the moment–rotation curves and corresponding skeleton curves are presented in Fig. 8.18. The initial imperfection size of the panel zone has a minor effect on configurations of the hysteretic properties and skeleton curves. It also has a negligible effect on the crack initiation, where differences of the instants at crack initiation are within a single displacement increment as shown in Fig. 8.18b.

8.5.2 Effect of Axial Load Ratio

In the experimental study, the axial load ratio, n, has not been comprehensively investigated due to limited number of specimens. Herein, specimen H1n3 with three different axial load ratios, i.e., 0.1, 0.3, and 0.5, was numerically investigated. The comparison results for the moment–rotation curves and corresponding skeleton curves are presented in Fig. 8.19, indicating that n has a significant effect on the strength of the connections. For the connection with a small n of 0.1, the load-carrying capacity increases gradually up to a large rotation of about 0.06 rad, while for the one with a large n of 0.5, the load-carrying capacity decreases gradually after the peak load is achieved at a relatively small rotation of about 0.025 rad. However, crack initiates at the same displacement amplitude level for all the three connections with different n, indicating that the axial load ratio has a small effect on the crack initiation.

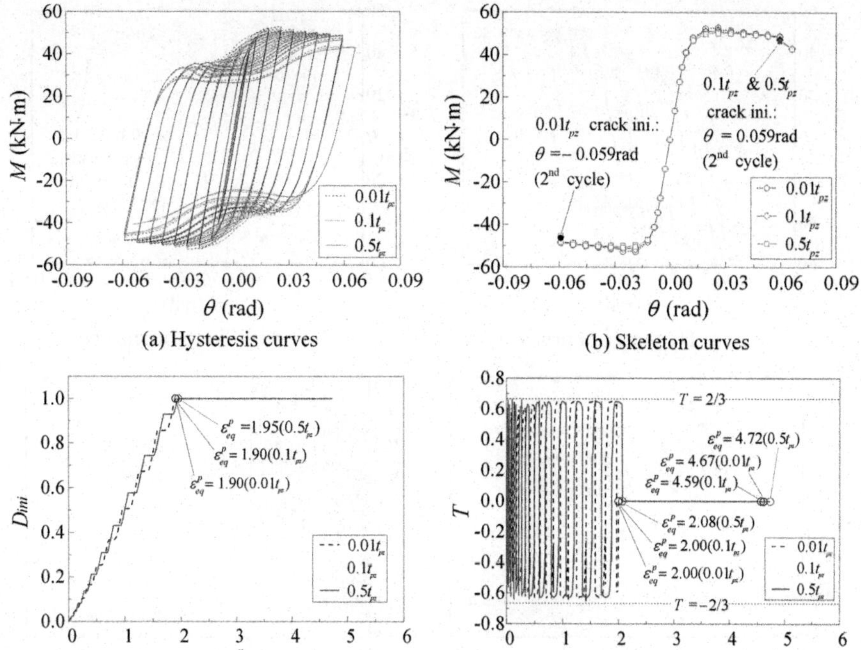

(a) Hysteresis curves (b) Skeleton curves

(c) Evolution of the damage index (d) Stress triaxiality histories

Fig. 8.18 Effect of initial geometrical imperfection

8.5.3 *Effect of Equivalent Panel Zone Width-to-Thickness Ratio*

Equivalent panel zone width-to-thickness ratio, λ_{pz}, has a great effect on the shear stability of the panel zone, which was also thought to have an effect on the post-buckling ductile fracture behavior of the specimens. Three connections with different λ_{pz}, i.e., 75, 90, and 105, that may lead to panel zone shear buckling were analyzed. The configurations of the analyzed connections are the same as that of H1n3 expect for the plate thickness of the panel zone. The comparison results for the normalized moment–rotation ($\bar{M}-\theta$) curves and corresponding skeleton curves are presented in Fig. 8.20, where M is normalized by the value of M when panel zone reaches its full plastic strength. Curves in Fig. 8.20 indicate that the strength is almost unaffected up to a large rotation of 0.06 rad, but the pinching effect of the hysteresis loops is slightly evident as λ_{pz} increases, and a decrease of the load-carrying capacity for the rotation beyond 0.06 rad becomes more rapid as λ_{pz} increases. The effect of λ_{pz} on crack initiation was also found to be negligible, which is similar to that of the initial geometric imperfection size. Figure 8.20b shows that macro-crack initiation becomes a bit earlier as the equivalent panel zone width-to-thickness ratio increases.

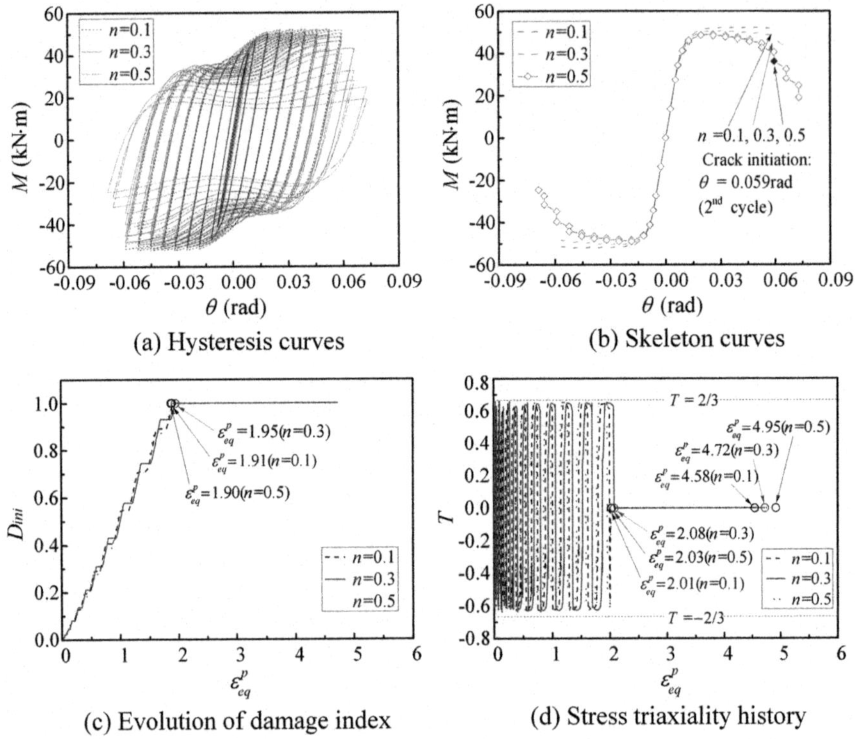

(a) Hysteresis curves (b) Skeleton curves

(c) Evolution of damage index (d) Stress triaxiality history

Fig. 8.19 Effect of axial load ratio

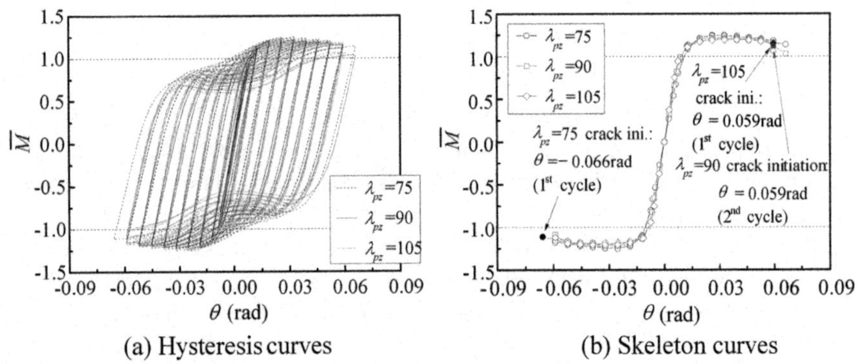

(a) Hysteresis curves (b) Skeleton curves

Fig. 8.20 Effect of equivalent panel zone width-to-thickness ratio

To further investigate the reason why the λ_{pz} has a minor effect on the cracking instants, local stress–strain states of the initially removed elements at each tensile loading half cycles were investigated for the models with different λ_{pz}. Figure 8.21a and b, respectively, show the histories of the incremental equivalent plastic strain,

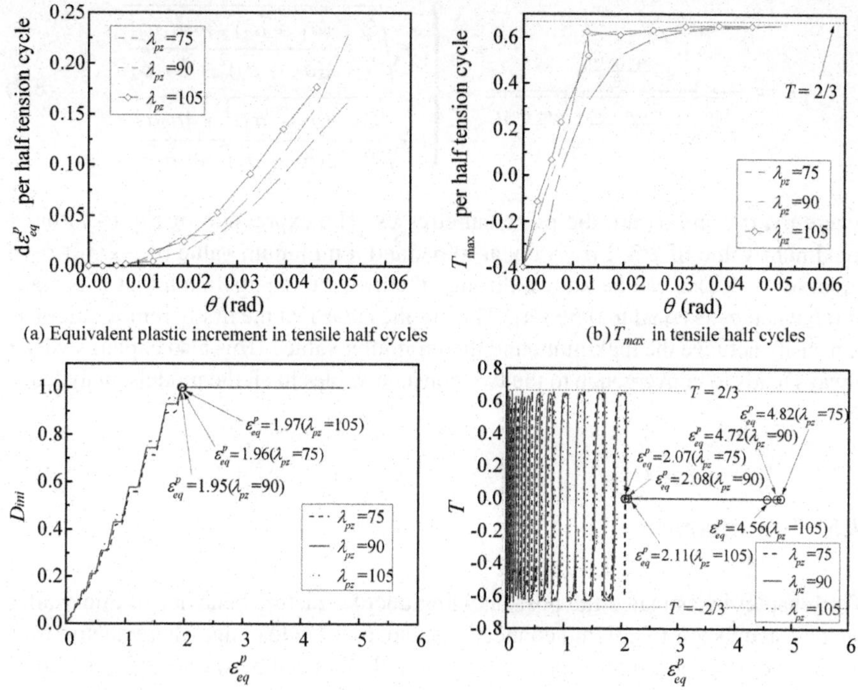

(a) Equivalent plastic increment in tensile half cycles (b) T_{max} in tensile half cycles

(c) Evolution of damage index (d) Stress triaxiality histories

Fig. 8.21 Effect of equivalent panel zone width-to-thickness ratio

$d\varepsilon_{eq}^{p}$, and the maximum stress triaxiality, T_{max}, within each tensile cycle, versus rotation of the specimens with different λ_{pz}. The figures show that $d\varepsilon_{eq}^{p}$ gets larger as the λ_{pz} increases and histories of T_{max} are similar for the three models. According to the fracture model employed in this study, a model with a larger λ_{pz} thus tends to crack earlier. However, Fig. 8.21c and d shows that the differences of the equivalent plastic strain till micro- and macro-crack initiation are all within 6% for the three models with different λ_{pz}. The differences in the number of loading cycles till macro-crack initiation for the three specimens are only within a single displacement level, which is unnoticeable from the viewpoint of global performance. The peak values of the positive and negative stress triaxiality at each half cycle are still respectively close to 2/3 and −2/3. To analyze the stress state of the firstly removed element, the three principal stresses were investigated. The thin-walled panel zone is in a plane stress state, where either the first or the third principal stress is equal to zero for the material at the firstly removed element. The stress triaxiality there can thus be expressed as,

$$T = \frac{\frac{\sigma_1+\sigma_2+\sigma_3}{3}}{\sqrt{\frac{(\sigma_1-\sigma_2)^2+(\sigma_1-\sigma_3)^2+(\sigma_2-\sigma_3)^2}{2}}} = \begin{cases} +\sqrt{\dfrac{2}{9}\cdot\dfrac{(\sigma_1-\sigma_2)^2+4\sigma_1\sigma_2}{2(\sigma_1-\sigma_2)^2+2\sigma_1\sigma_2}} \\ -\sqrt{\dfrac{2}{9}\cdot\dfrac{(\sigma_2-\sigma_3)^2+4\sigma_2\sigma_3}{2(\sigma_2-\sigma_3)^2+2\sigma_2\sigma_3}} \end{cases} \quad (8.7)$$

where σ_1, σ_2, and σ_3 are the principal stresses. The expression of Eq. (8.7) has a maximum value of 2/3 if σ_1 is equal to σ_2, and a minimum value of −2/3 if σ_2 is equal to σ_3. It also can be known through mathematic manipulation that T is equal to 0.6 when σ_2 is equal to 0.563 σ_1. This implies that T at the firstly remove element can easily achieve the maximum and the minimum values. This also explains why T always tends to convergence to the two constant values in all the models analyzed in this study.

8.6 Summaries

This chapter investigated the post-buckling ductile fracture behavior of thin-walled welded steel beam-to-column connections under ULCF loading, where ductile fracture of the base metal at the panel zone is mainly induced by severe shear buckling. Experimental study on two full-scale thin-walled welded steel beam-to-column connections was conducted. The post-buckling ductile cracking process was simulated using a proposed micro-mechanism-based ductile fracture model and further investigated on effects of initial geometrical imperfection, axial load ratio and equivalent panel zone width-to-thickness ratio, on crack initiation and propagation of the connections using the validated cyclic ductile fracture model. Based on the experimental and numerical studies on shear buckled panel zones, the following conclusions can be drawn:

- The thin-walled welded steel beam-to-column connections have a large rotation capacity far beyond 0.03 rad, indicating favorable energy dissipation capacity of the connections if the adjacent welds are properly handled.
- The adopted cyclic ductile fracture model with the cyclic plasticity model, i.e., the Chaboche model, can well capture the hysteretic properties and crack initiation behavior of the connections under ULCF loading.
- Damage of the investigated connections is mainly concentrated on X-shaped bands within the panel zone, and also at the weld toes of the perimeter of the panel zone. Damage at the concave side is more remarkable than the convex one.
- Initial imperfection size is found to have minor effect on the hysteretic properties and crack initiation of the connections.
- The axial load ratio has a great effect on the hysteretic properties of the connections, while has minor effect on the crack initiation behaviors.
- The equivalent panel zone width-to-thickness ratio is found to have minor effect on the hysteretic properties for rotation below around 0.06 rad, and the load-

carrying capacity decreases a bit more rapidly as the equivalent panel zone width-to-thickness ratio increases. The crack initiation of the connections is also almost unaffected by the equivalent panel zone width-to-thickness ratio.

- The plane stress state of the thin-walled panel zone leads to the fact that the maximum and minimum stress triaxialities, respectively, approach 2/3 and −2/3.

It should be noted that the crack propagation is simulated by the simple element deletion method, where closure and interaction of cracks under ULCF loading cannot be realistically reproduced. If the closure and interaction of cracks are considered, the loads may increase a bit owing to the contact forces. The crack propagation process can also be a bit different due to change of stress states at the crack tips when cracks are closed and interact with each other.

References

ABAQUS (2010) ABAQUS standard manual (version 6.10). Karlsson & Sorensen Inc., Hibbitt, Pawtucket

ANSI/AISC341-10 (2010) Seismic provisions for structural steel buildings. American Institute of Steel Construction, Chicago

ANSI/AISC360-10 (2010) Specification for structural steel buildings. American Institute of Steel Construction, Chicago

Bao Y, Wierzbicki T (2005) On the cut-off value of negative triaxiality for fracture. Eng Fract Mech 72:1049–1069

GB50011-2010 (2010) Code for seismic design of buildings. Architecture & Building Press, Beijing

Gurson AL (1975) Continuum theory of ductile rupture by void nucleation and growth. Part I. Yield criteria and flow rules for porous ductile media. Division of Engineering, Brown University, Providence

Jia L-J, Kuwamura H (2014) Ductile fracture simulation of structural steels under monotonic tension. J Struct Eng (ASCE) 140:04013115

Jia L-J, Kuwamura H (2015) Ductile fracture model for structural steel under cyclic large strain loading. J Constr Steel Res 106:110–121

Jia L-J et al (2014) Experimental and numerical study of postbuckling ductile fracture of heat-treated SHS stub columns. J Struct Eng (ASCE) 140:04014044

Jia L-J et al (2016a) Experimental and numerical study on ductile fracture of structural steels under combined shear and tension. J Bridge Eng (ASCE) 04016008

Jia L-J et al (2016b) Ductile crack initiation and propagation of structural steels under cyclic combined shear and normal stress loading. Constr Build Mater 112:69–83

Kanvinde A, Deierlein G (2006) The void growth model and the stress modified critical strain model to predict ductile fracture in structural steels. J Struct Eng (ASCE) 132:1907–1918

Kanvinde A, Deierlein G (2007) Cyclic void growth model to assess ductile fracture initiation in structural steels due to ultra low cycle fatigue. J Eng Mech (ASCE) 133:701–712

Kanvinde A, Deierlein G (2008) Validation of cyclic void growth model for fracture initiation in blunt notch and dogbone steel specimens. J Struct Eng (ASCE) 134:1528–1537

Krawinkler H (1978) Shear in beam-column joints in seismic design of steel frames. Eng J 15:82–91

Krawinkler H et al (1975) Shear behavior of steel frame joints. ASCE J Struct Div 101:2317–2336

Liao F et al (2015) Ductile fracture prediction for welded steel connections under monotonic loading based on micromechanical fracture criteria. Eng Struct 94:16–28

McClintock FA (1968) A criterion for ductile fracture by the growth of holes. J Appl Mech 35:363–371

Pan L et al (2016) Experimental evaluation of the effect of vertical connecting plates on panel zone shear stability. Thin Wall Struct 99:119–131
Panontin TL, Sheppard SD (1995) The relationship between constraint and ductile fracture initiation as defined by micromechanical analyses. ASTM STP 1256, ASTM, West Conshohoken, pp 54–85
Popov EP et al (1985) Cyclic behavior of large beam-column assemblies. Earthq Spectra 1:203–238
Qian X et al (2005) Simulation of ductile fracture of circular hollow section joints using the Gurson model. J Struct Eng (ASCE) 131:768–780
Rice JR, Tracey DM (1969) On the ductile enlargement of voids in triaxial stress fields. J Mech Phys Solids 17:201–217
Rousselier G (1987) Ductile fracture models and their potential in local approach of fracture. Nucl Eng Des 105:97–111
Tvergaard V (1981) Influence of voids on shear band instabilities under plane strain conditions. Int J Fract 17:389–407
Tvergaard V, Needleman A (1984) Analysis of the cup-cone fracture in a round tensile bar. Acta Metall 32:157–169
Zhou F et al (2015) Dependence of the cyclic response of structural steel on loading history under large inelastic strains. J Constr Steel Res 104:64–73

Chapter 9
Cyclic Plasticity of Aluminum in Large Plastic Strain Ranges

Abstract Aluminum has been increasingly employed in space, building, and other structures owing to its light weight and high durability. Compared with structural steel, aluminum has relatively low ductility and is more apt to fail during strong earthquakes. Seismic loading is associated with random strain amplitudes, which makes it necessary to calibrate a plasticity model at the full strain range. Based on the achievements in the previous chapter, this chapter aims to propose a straightforward approach to accurately evaluate hysteretic properties of aluminum material and structures under variable-amplitude cyclic loading within the full strain range till fracture, where only representative mechanical variables such as yield strength and tensile strength are required to calibrate the generalized Armstrong–Frederick model. The newly proposed method is validated at both material and member levels, respectively, through quasi-static cyclic experiments on double-edge-notched specimens and aluminum buckling-restrained braces. The validation results show that the proposed method can well describe cyclic plasticity of aluminum members at the full strain range.

9.1 Introduction

Compared with steel, aluminum has high corrosion resistance and thus less maintenance expense, which makes it suitable for structures in harsh environments such as bridges by the sea. There are a number of applications of aluminum structures in bridge engineering (e.g., Dey et al. 2016). Aluminum also has a relatively high strength–mass ratio, leading to light weight of the structures and also more economical foundations. Light self-weight of aluminum structures leads to less transportation cost and lower capacity requirement for the hoisting equipment. This favorable property also makes structural aluminum competitive for large-span roof structures as illustrated in Fig. 9.1.

A number of experimental and numerical studies on aluminum members and connections have been conducted (e.g., Dørum et al. 2010; De Matteis et al. 2000, 2008; Khadyko et al. 2015; Matteis et al. 2001; Mazzolani et al. 2011; Moen et al. 1999;

L.-J. Jia and H. Ge, *Ultra-low-Cycle Fatigue Failure of Metal Structures under Strong Earthquakes*, Springer Tracts in Civil Engineering,
https://doi.org/10.1007/978-981-13-2661-5_9

(a) Nanjing Usnisa Palace (b) Shanghai Chenshan Botanical Garden

Fig. 9.1 Applications of aluminum in structural engineering

Saleem et al. 2012; Su et al. 2015; Su et al. 2016). However, studies on cyclic plasticity of structural aluminum are still very limited (e.g., Yin et al. 2004), especially for cases under variable-amplitude cyclic loading within extremely large plastic strain range (Liu et al. 2017; Rosien and Ostertag 2009). In regions with a high seismic risk, it is of great importance to investigate the behaviors of structures under cyclic large plastic strain loading, particularly for aluminum structures, since ductility of aluminum is much lower than that of structural steel (Jia and Kuwamura 2014a). Elongation of structural aluminum is about 12%, while that of mild steel is commonly larger than 25%. This implies that aluminum structures are more vulnerable to fracture under seismic loading. Aluminum dampers are also proposed by some researchers (Brando et al. 2013), where evaluation of the hysteretic properties is closely related with cyclic plasticity under extremely large plastic strain loading.

Cyclic plasticity of metal structures has been extensively investigated (e.g., Chang and Lee 1986a, b; Hu et al. 2016a, b; Shen et al. 1995; Ucak and Tsopelas 2012; Wang et al. 2016). However, most of them are focused on cyclic plasticity in small strain ranges, e.g., within 5% (Ucak and Tsopelas 2011). During a strong earthquake, structural members can sustain large plastic strain as much as the fracture strain, which can, respectively, exceed 100 and 40% for structural steel and aluminum, especially for ductile fracture and ultra-low-cycle fatigue problems under monotonic or cyclic large plastic strain loading (Ge and Kang 2014; Jia et al. 2016a, b; Jia and Kuwamura 2015; Kang et al. 2015; Liao et al. 2015; Tabatabai 2010).

In the previous study, a method to evaluate hysteretic properties of mild steel using only tension coupon test results was proposed (Jia and Kuwamura 2014b). However, this method is still limited in practice, since structural engineers can obtain only very limited information about material properties. In such cases, one has to calibrate plasticity model parameters with the limited known mechanical properties. This study aims to propose a simple approach to calibrate plasticity model parameters with the minimum number of mechanical variables, such as yield strength and tensile strength. In this study, the generalized Armstrong–Frederick (A-F) model is employed (Frederick and Armstrong 2007), which is also termed as the Chaboche model. First, the Chaboche model and the corresponding calibration method are briefly introduced. Then, experimental study on structural aluminum material under different variable-amplitude cyclic loadings up to fracture was conducted. Corresponding numerical simulation with the calibrated parameters obtained by the newly proposed method was conducted. Validity of the proposed method was verified at the material level through comparison between the experimental and numerical results. Finally, the proposed method was further validated at the member level through comparison of experimental and numerical results of aluminum buckling-restrained braces (BRBs), which are employed to dissipate seismic energy through cyclic plastic deformation. The above experimental and numerical results indicate that the newly proposed method and corresponding calibrated model parameters can well evaluate the hysteretic properties of structural aluminum at the full strain range. Owing to simplicity and accuracy of the newly proposed method, there is a great potential for the method to be applied to evaluating cyclic plasticity of metal structures in a variety of engineering fields. The calibrated plasticity model parameters can also be employed to evaluate seismic performance of aluminum structures under cyclic variable-amplitude loading.

9.2 Approach to Calibrate Plasticity Model Parameters Using Minimum Number of Mechanical Variables

In the previous study of the authors, it has been found that parameters of a cyclic plasticity model such as the Chaboche model with IH can be calibrated using only tension coupon test results. This is mainly owing to the observation of fatigue tests in the literature (Kuhlmann-Wilsdorf and Laird 1979). Cyclic plasticity of metals can thus be straightforwardly correlated with the corresponding monotonic coupon test results. In the previous study on cyclic plasticity of structural steel, it has been found that the IH and KH components approximately each take half the hardening stress in a monotonic tension coupon test (Jia and Kuwamura 2014b).

For the uniaxial stress state, the incremental formation of the backstress in the Chaboche model with IH can be expressed as,

$$\boldsymbol{\alpha} = \begin{cases} \frac{C_0}{\gamma}[1 - \mathbf{exp}(-\gamma \boldsymbol{\varepsilon}_p)] & \text{when } \gamma \neq 0 \\ C_0 \boldsymbol{\varepsilon}_\mathbf{p} & \text{when } \gamma = 0 \end{cases} \tag{9.1}$$

Equation (9.1) indicates that the KH component, i.e., the backstress, has a limiting value of C_0/γ when γ is not equal to zero.

For a uniaxial stress state, the IH component can be expressed as,

$$R = Q_\infty[1 - \mathbf{exp}(-k \cdot \varepsilon_{eq})] \tag{9.2}$$

Likewise, the IH component also has a limiting value of Q_∞. For the Chaboche model with IH, one has to determine the initial yield stress, σ_{y0}, KH-related param-

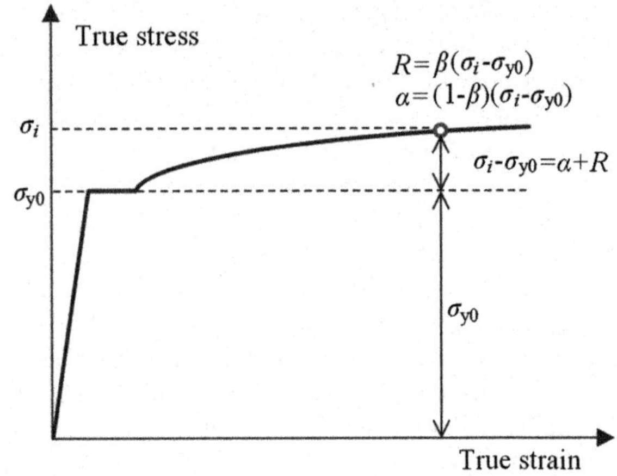

Fig. 9.2 Proportion of isotropic and kinematic hardening in tension coupon test

eters, i.e., C_i and γ_i, and IH-related parameters, i.e., Q_∞ and k. The total hardening stress equaling to $R + \alpha$ can be readily obtained based on the following equation,

$$\sigma_i = \sigma_{y0} + R + \alpha \quad (9.3)$$

where σ_i is the true stress obtained from a tension coupon test. By postulating that the proportion of the IH component in the total hardening strength equals to β, then R and α can be obtained as illustrated in Fig. 9.2,

$$R = \beta(\sigma_i - \sigma_{y0}) \quad (9.4)$$

$$\alpha = (1 - \beta)(\sigma_i - \sigma_{y0}) \quad (9.5)$$

The value of β can be obtained based on comparison of monotonic and cyclic test results, which can be taken as a material constant for metal with the same grade. Once β is determined, R versus true strain data can be obtained according to Eq. (9.4), and Q_∞ and k can thus be obtained based on fitting. Likewise, α versus true strain data can also be obtained according to Eq. (9.5), and C_i and γ_i can also be simply obtained based on regression analysis.

For the Chaboche model with IH, one has to determine the following model parameters, C_1, γ_1, C_2, Q_∞, and k. According to Eqs. (9.4) and (9.5), once the value of β is determined, both the IH- and KH-related parameters can be readily determined through regression analysis. In this chapter, two backstresses were employed for the KH components for structural aluminum. The main reason why two backstresses were employed is for its simplicity to correlate the model parameters with the mechanical properties such as yield stress, necking initiation strain, and tensile strength. For the KH component, one backstress with an exponential function was employed, and the other one a linear function; i.e., γ_2 was taken as zero. The parameter C_2 is correlated with the hardening rate of the linear part of the KH component, which is commonly employed to describe the hardening behavior at extremely large plastic strain ranges. At these strain ranges, the hardening rate can be assumed as a linear relationship.

To ensure that ultimate strength of structural components is accurately evaluated, tensile strength of a material in the cyclic plasticity model should be consistent with that of the monotonic tension coupon test results during the calibration process. Commonly, tensile strength of a material corresponds to the instant of the necking initiation. Thus, sum of the KH and IH stresses at the instant of necking initiation of a tension coupon should equal to the hardening stress at the instant of necking initiation of a material. The following equation can be obtained,

$$\sigma_{\text{neck}} - \sigma_{y0} = R_{\text{neck}} + \alpha_{\text{neck}} \tag{9.6}$$

where σ_{neck} = true stress at the instant of necking initiation; R_{neck} = IH component at the instant of necking initiation; α_{neck} = KH component at the instant of necking initiation. σ_{neck} can be obtained from the following equation,

$$\sigma_{\text{neck}} = s_{\text{neck}} \cdot (1 + e_{\text{neck}}) \tag{9.7}$$

where s_{neck} = engineering stress at the instant of necking initiation, i.e., the tensile strength of a material; e_{neck} = engineering strain at the instant of necking initiation. R_{neck} can be obtained according to Eqs. (9.2) and (9.4),

$$R_{\text{neck}} = \beta \cdot (\sigma_{\text{neck}} - \sigma_{y0}) = Q_{\infty} \cdot [1 - \mathbf{exp}(-k \cdot \varepsilon_{\text{neck}})] \tag{9.8}$$

where $\varepsilon_{\text{neck}}$ = true strain at the instant of necking initiation. α_{neck} can be obtained based on Eqs. (9.1) and (9.5) as follows,

$$\alpha_{\text{neck}} = (1 - \beta) \cdot (\sigma_{\text{neck}} - \sigma_{y0}) = \frac{C_1}{\gamma_1} \cdot [1 - \mathbf{exp}(-\gamma_1 \cdot \varepsilon_{\text{neck}})] + C_2 \cdot \varepsilon_{\text{neck}} \tag{9.9}$$

Commonly, β, $\varepsilon_{\text{neck}}$, C_2, γ_1, and k can be taken as material constants which have minor deviation for metals with the same grade. Consequently, parameters of the Chaboche model with IH that have to be determined are C_1 and Q_{∞}. For a ductile metal such as aluminum, it is postulated that β, $\varepsilon_{\text{neck}}$, C_2, γ_1, and k are known, and one can readily obtain the values of C_1 and Q_{∞} according to Eqs. (9.7)–(9.9) once the yield strength, σ_{y0}, and tensile strength, s_{neck}, are obtained.

In this study, the proposed calibration method was firstly illustrated using experimental and numerical studies on aluminum double-edge-notched (DEN) specimens, where the parameters $\varepsilon_{\text{neck}}$, C_2, γ_1, and k are obtained using the corresponding tension coupon test results. Subsequently, the newly proposed method was illustrated at the structural member level using test results of aluminum BRBs, where only the yield strength and the tensile strength were employed during the calibration process, since the values of $\varepsilon_{\text{neck}}$, C_2, γ_1, and k have been determined through the experimental and analysis results of the DEN specimens.

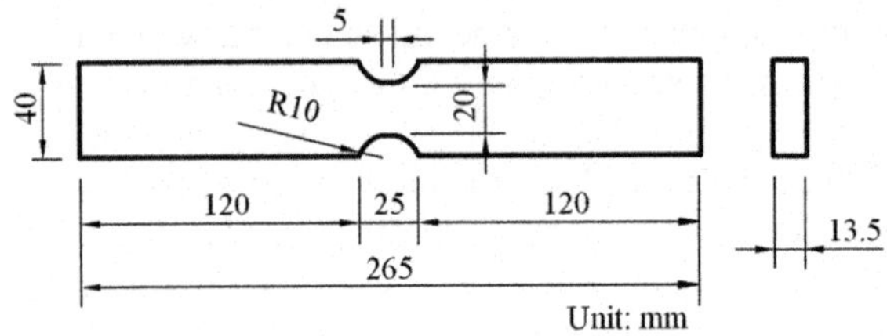

Fig. 9.3 Configuration of double-edge-notched specimen

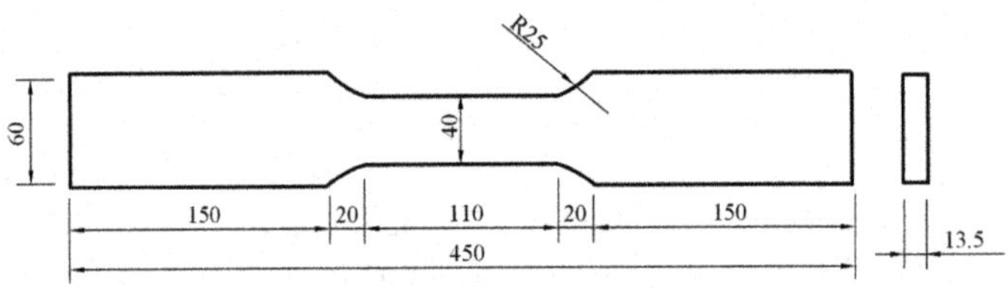

Fig. 9.4 Configuration of tension coupon test

9.3 Verification of the Proposed Method at Material Level

9.3.1 Experimental Study on Aluminum DEN Specimens

Eight aluminum DEN specimens with the configuration shown in Fig. 9.3 were designed, where the central segment with a small length of 5 mm has the same cross-sectional area to produce stress–strain state as uniform as possible. The compact configuration was determined for the avoidance of premature buckling under compression. All the specimens were manufactured from the same aluminum plate with a nominal thickness of 13.5 mm, which was to exclude the influence of the material deviation. The material grade is 6061-T6, where the mechanical properties and chemical composition are given in Table 9.1. Tension tests on four coupons with the configuration shown in Fig. 9.4 were conducted, and the results were employed to calibrate the model parameters of the Chaboche model with IH utilizing the aforementioned method.

All the tests were conducted under room temperature at a quasi-static speed utilizing an MTS system shown in Fig. 9.5, and the displacement and load capacities of the loading system are, respectively, ±75 mm and 250 kN. An extensometer with a gage length of 50 mm was employed to measure the deformation within the gage length. The two ends of a DEN specimen were clamped by the loading heads. The bottom loading head is fixed, and the top one is movable. All the tests were conducted under displacement control using the data of the extensometer.

To investigate cyclic plasticity of structural aluminum at various strain ranges, totally eight different loading histories shown in Fig. 9.6 were designed. The first two specimens were, respectively, loaded under monotonic tension and compression as

Table 9.1 Mechanical properties and chemical composition of aluminum 6061-T6

Material	Mechanical properties				Chemical composition (weight %)								
	Yield stress (MPa)	Tensile strength (MPa)	Elongation (%)	$\varepsilon_{\text{Neck}}$ (%)	Al	Mg	Si	Fe	Cu	Mn	Zn	Cr	Ti
DEN	257	301	12	11	98.0	1.0	0.6	0.1	0.2	0.03	0.01	0.09	0.01
BRB	294	331	12	/	/	/	/	/	/	/	/	/	/

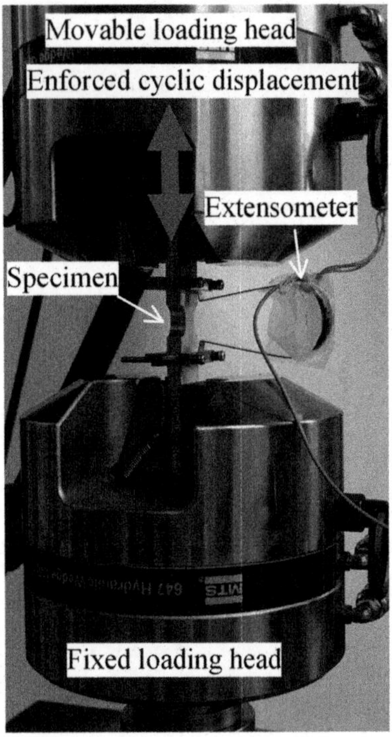

Fig. 9.5 Test setup of double-edge-notched specimens

illustrated in Fig. 9.6a and b. The two specimens were employed to capture the instants of necking initiation and buckling, and the other loading histories were designed based on the experimental results of these two specimens. The third specimen was tested under a single-full-cycle loading shown in Fig. 9.6c, where the plastic strain amplitude is relatively small. The fourth specimen was tested under five-constant-amplitude-cycle loading and then pulled to rupture shown in Fig. 9.6d, which was to investigate the stabilization effect of the cyclic hardening. The fifth specimen was tested under pre-necking cyclic loading with two constant-amplitude cycles shown in Fig. 9.6e, where the necking initiation instant is also marked in the curve. Similarly, Fig. 9.6f gives the post-necking cyclic loading history. The last two loading histories shown in Fig. 9.6g and h were employed to validate the proposed calibration method at the full strain range, i.e., both pre- and post-necking ranges. The former is an incremental loading history at the full strain range up to fracture, and the latter has two constant-amplitude loading cycles at both the pre- and post-necking stages. Through the aforementioned loading histories, plasticity behaviors of structural aluminum under cyclic extremely large plastic strain loading can be comprehensively evaluated.

Coupon test results given in Table 9.1 indicate that the tensile-strength-to-yield-strength ratio of the aluminum 6061-T6 is 1.17. Macro-fracture surfaces shown in Fig. 9.7 indicate that all the DEN specimens failed in the ductile fracture mode for the cyclic large plastic strain loading histories employed in this study. The load-

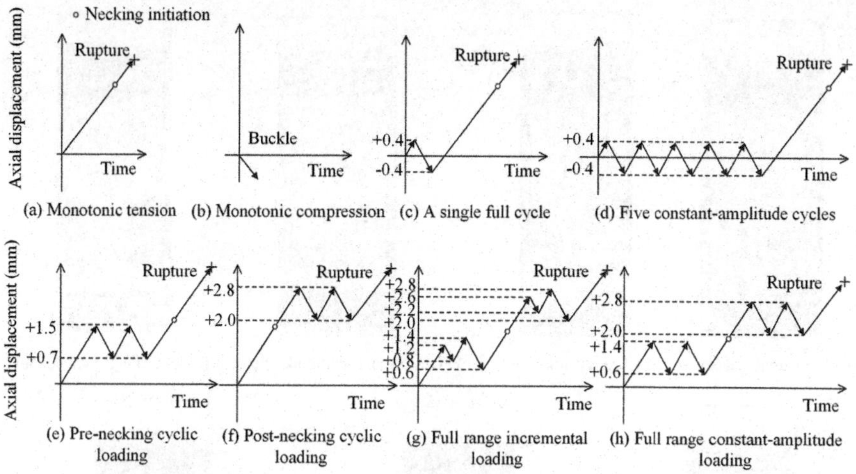

Fig. 9.6 Loading protocols for double-edge-notched specimens

–displacement curves are given in Fig. 9.8, where the instants at the peak load and rupture are also given in the figure. The nominal initial yield force calculated from the coupon test results is 70 kN. The peak load ranges from 85.8–89.5 kN, and the differences are within 4% for the DEN specimens under different loading histories. The displacement at instant of the peak load also differs from each other, where no clear tendency can be found. The loading history is found to have remarkable effect on the rupture displacement, and the maximum deviation is 27%, while that of the rupture load is 6%. Comparison of the pre- and post-necking cyclic loading cases, respectively, shown in Fig. 9.8d and e implies that reversals at the post-necking stage can lead to reduced rupture displacement; i.e., reversals with the same amplitude at extremely large plastic strain ranges can result in more damage to the aluminum. This is mainly due to the uneven strain distribution over the minimum cross section after necking.

9.3.2 Numerical Simulation of Aluminum DEN Specimens

Numerical analyses using three-dimensional solid element models were conducted for the DEN specimens using commercial finite element (FE) software (ABAQUS 2010). The elements *C3D8R* were employed considering both accuracy and efficiency. For ease of the numerical simulation, a partial model shown in Fig. 9.9b is preferred, since the experiments were controlled using the displacement data within the gage length of the extensometer. The elements within the uniform cross-sectional part were finer than those in the other parts to capture the strain concentration there. A comparison between the analysis results of the whole and the partial models shown

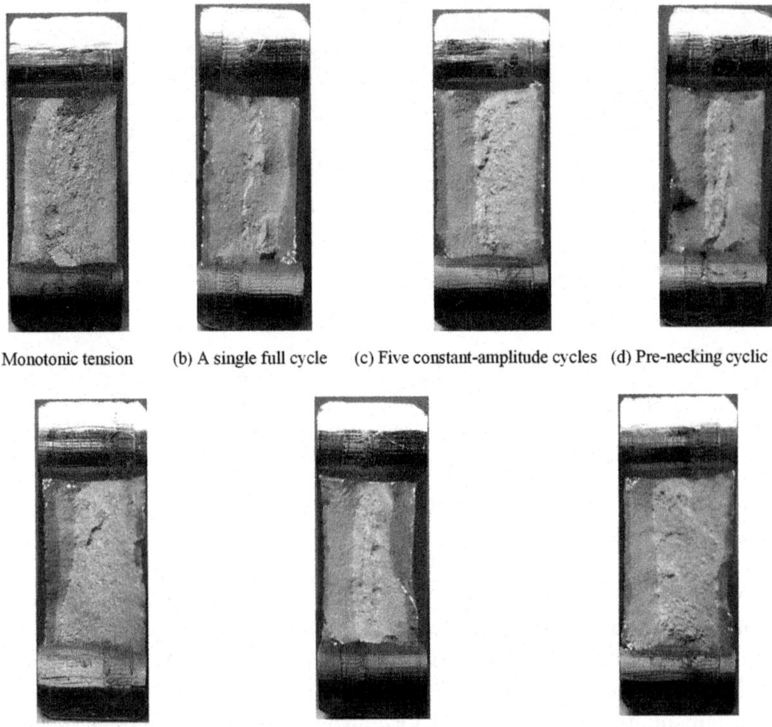

(a) Monotonic tension (b) A single full cycle (c) Five constant-amplitude cycles (d) Pre-necking cyclic loading

(e) Post-necking cyclic loading (f) Full range incremental loading (g) Full range constant-amplitude loading

Fig. 9.7 Fracture surfaces of DEN specimens

in Fig. 9.9 was carried out, and the result shows that the two models give almost the same results. The partial model with boundary conditions illustrated in Fig. 9.10 was thus employed in the analyses for DEN specimens. It should be noted that the horizontal freedom of the top and bottom ends is free, and enforced displacement the same as that of the experiment was applied to the top end.

9.3.3 Calibration of Plasticity Model Parameters for Aluminum DEN Specimens

Regression analyses based on the monotonic tension coupon test results were conducted to determine γ_1, C_2, and k, since no reference data can be employed. The determined values of the parameters together with $\varepsilon_{\text{neck}}$ can be taken as material constants and further employed in the analysis of the aluminum BRBs. The values of γ_1, C_2, and k for aluminum 6061-T6 are given in Table 9.2, which are, respectively, equal to 18.5, 25 MPa, and 14.1. Since it has been found that β is close to 0.5 for

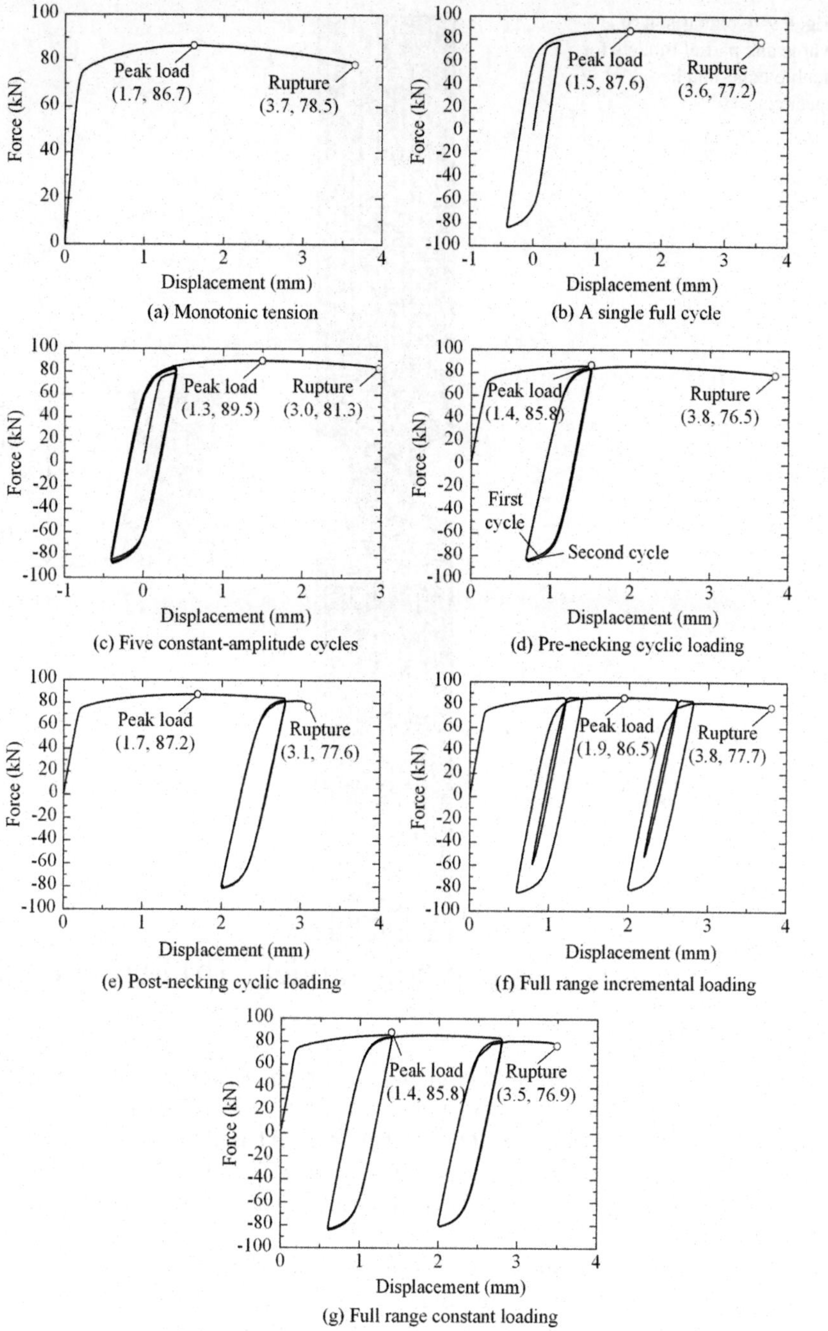

Fig. 9.8 Experimental results of DEN specimens

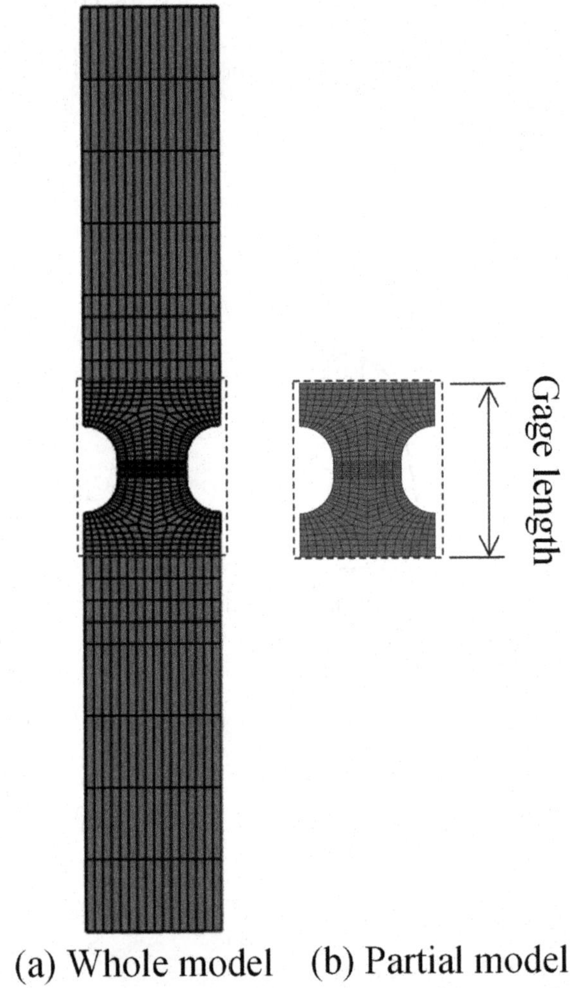

Fig. 9.9 Comparison of whole and partial models for double-edge-notched specimens

most ductile metals such as mild steel and aluminum, a value ranging from 0.0 to 1.0 was selected in this study to investigate its optimal value for structural aluminum 6061-T6. For different β, C_1 and Q_∞ can be obtained based on the aforementioned calibration method, and their values are also given in Table 9.2.

Contour plots of the equivalent plastic strain at the instant of rupture obtained from the numerical results are given in Fig. 9.11, indicating great strain concentration at the mid-thickness of the minimum cross section. It can be found from Fig. 9.11 that the equivalent plastic strain at the instant of rupture ranges from 36 to 64%, which is fairly large compared with conventional metal plasticity problems. The equivalent plastic strain increases drastically for the post-necking stage and also increases as the number of reversals increases, e.g., the five-constant-amplitude-cycle loading.

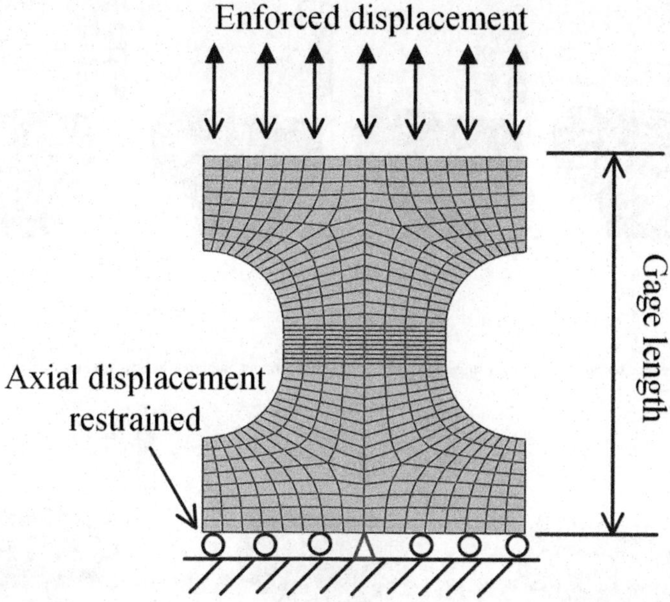

Fig. 9.10 FE model and boundary conditions for double-edge-notched specimens

Table 9.2 Calibrated model parameters of plasticity model

Specimen	β	C_1 (MPa)	Q_∞ (MPa)	σ_{y0} (MPa)	C_2 (MPa)	γ_1	k	γ_2
DEN	0	1634	0	257	25	18.5	14.1	0
	0.4	922	39					
	0.5	759	49					
	0.6	596	59					
	1	0	94					
BRB	0	1551	0	294				
	0.4	872	37					
	0.5	717	46					
	0.6	562	56					
	1	0	89					

Comparisons between the experimental and numerical results for the load–displacement curves of the DEN specimens were given in Fig. 9.12, where the one under monotonic compression is not shown due to premature buckling. From the figure, the following observations can be found:

(1) With the proposed calibration method, the Chaboche model with IH can evaluate both monotonic tension and cyclic plasticity of structural aluminum 6061-T6 with good accuracy for β ranging from 0.4 to 0.6.
(2) For small plastic strain ranges shown in Fig. 9.12b and c, a β of 0.4 gives the most accurate evaluation.

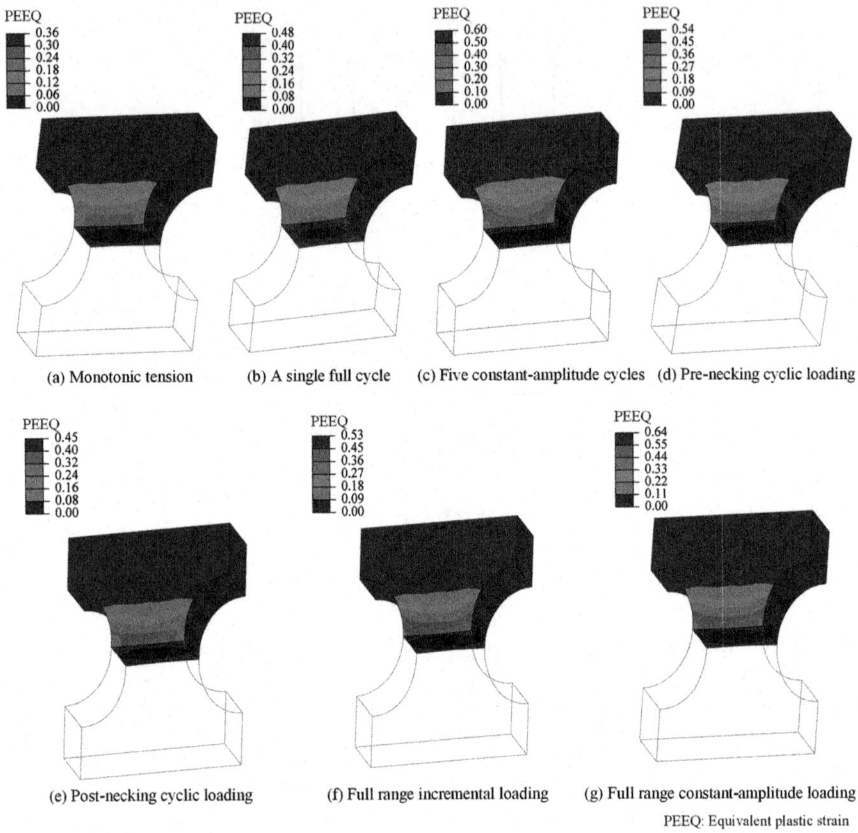

Fig. 9.11 Contour plots of equivalent plastic strain at instant of rupture

(3) Without a memory surface, the Chaboche model with IH gives an overestimation of the stress at subsequent constant-amplitude loading cycles as shown in Fig. 9.12c.
(4) For the medium strain ranges shown in Fig. 9.12d, i.e., the pre-necking constant-amplitude loading case, a β of 0.5 gives the best evaluation result at the tension side, while a β of 0.6 for the compression side.
(5) For the medium and large plastic strain ranges shown in Fig. 9.12d–g, i.e., the pre- and post-necking loading cases, a β of 0.5 gives the best evaluation result at the tension side, while a β of 0.6 for the compression side.
(6) For all the cyclic loading cases, the Chaboche model with IH overestimates the stress at the elastic–plastic transition regions using the proposed method. This is mainly due to the fact that the hardening rate of a monotonic tension test at the elastic–plastic transition region is higher than the one of a post-yielding loading cycle in a cyclic loading case (Yoshida and Uemori 2002).

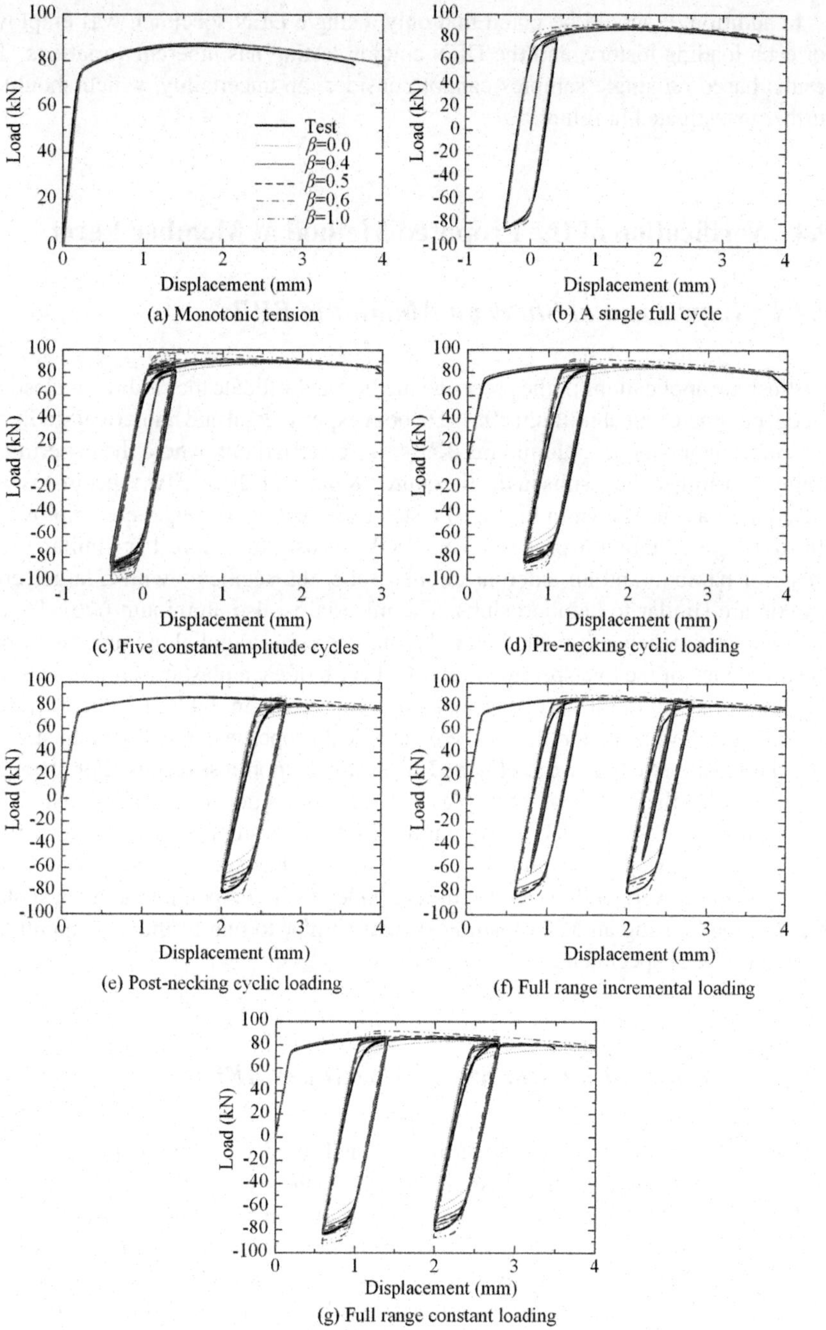

Fig. 9.12 Comparison of experimental and numerical results for DEN specimens

In addition, it should be noted that only a single DEN specimen was employed for each loading history, and the DEN coupon testing has inherent variations. The results based on single samples cannot consider the uncertainty, which should be further investigated in future.

9.4 Verification of the Proposed Method at Member Level

9.4.1 Experimental Study on Aluminum BRBs

To illustrate applications of the proposed method and validate the calibrated plasticity model parameters of aluminum 6061-T6, both experimental and numerical studies on structural members, i.e., aluminum BRBs, were carried out, where the experiments were according to the results in the literature (Wang et al. 2018). Two specimens with the configurations shown in Fig. 9.13 tested under cyclic loading were employed to illustrate the calibration process. The BRB consists of a core bar similar to the shape of bamboo and an external circular tube, the segments with a larger cross section are similar to bamboo slubs. The material is also aluminum 6061-T6, and the two specimens were manufactured from the same round aluminum bar, where the difference of the two specimens are the length of each plastic straining segment. The chemical composition of the material is not given in Table 9.2 due to lack of the mill sheet of the material. For the mechanical properties, only the yield strength and the tensile strength were obtained from the coupon test results. For specimen S2-L4S20G1, the length of each plastic straining portion in each segment is 40, and 60 mm for S2-L6S20G1 as shown in Fig. 9.13. A stopper at mid-length of the BRB is to avoid rigid movement between the core bar and the external tube. The two specimens were both tested to rupture under room temperature at a quasi-static loading speed using an MTS loading system similar to that of the aforementioned testing on DEN specimens.

9.4.2 Numerical Simulation of Aluminum BRBs

Numerical simulations using three-dimensional solid element models were conducted for the aluminum BRBs using ABAQUS, where the element type is the same as that of the DEN specimens. An FE model shown in Fig. 9.14 was established in the implicit analysis module. As shown in the figure, the external tube was fixed, and cyclic displacement loading was applied to two ends of the core bar. The core bar was meshed with fine elements to accurately capture the hysteretic properties and deformed shapes, and the mesh of external restraining tube could be relatively coarse owing to small stress in the component.

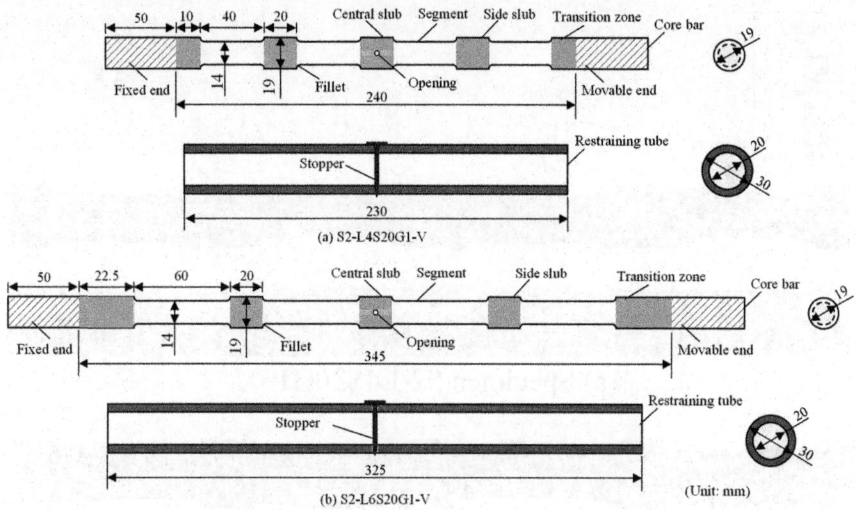

Fig. 9.13 Configurations of aluminum BRBs

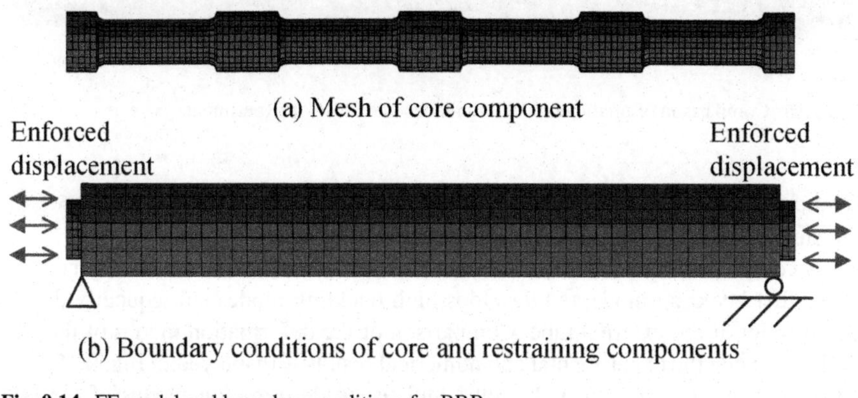

Fig. 9.14 FE model and boundary conditions for BRBs

9.4.3 Calibration of Plasticity Model Parameters Using Only Typical Mechanical Properties

Two coupon tests were conducted to obtain the required variables during calibration process of the plasticity model. Only the typical mechanical properties obtained from the coupon tests were employed during calibration of the Chaboche model with IH, since β, $\varepsilon_{\text{neck}}$, γ_1, and k have been postulated to be the same as those of the DEN specimens. This is a bit different from the calibration process of the DEN specimens, and the minimum number of mechanical properties is required, i.e., the yield strength and the tensile strength. Based on the newly proposed calibration method, the parameters of the Chaboche model with IH for the aluminum BRBs can

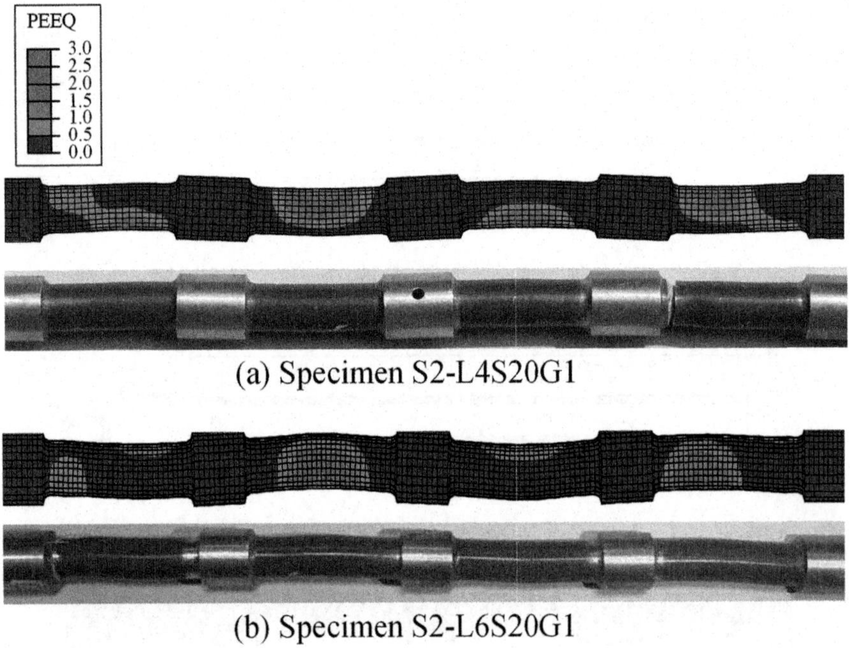

Fig. 9.15 Comparison of predicted buckling modes and those of experiments

be obtained, which are listed in Table 9.2. Likewise, β ranging from 0.4 to 0.6 were selected for the calibration of the plasticity model parameters.

As compressive deformation increases, the core bar of the BRB initially buckles in the first buckling mode and develops high buckling modes subsequently due to the restraint of the external tube. Comparison of the deformation modes of the two BRBs between the experimental and numerical results is displayed in Fig. 9.15. The numerical simulations can both well simulate the deformation modes of the two specimens. Meanwhile, it can also be found that the accumulated equivalent plastic strain can reach as much as about 300%, which is fairly large compared with other engineering fields such as mechanical engineering.

The load–displacement curves of the experiments and the simulations are, respectively, compared in Figs. 9.16 and 9.17 for specimens S2-L4S20G1 and S2-L6S20G1. The comparison results indicate that acceptable accuracy can be achieved using the calibrated plasticity model parameters obtained by the proposed calibration method. It can also be found that a β of 0.5 can well capture both the tensile and compressive peak loads. The ones with a β of 0.4 underestimate the peak loads and of 0.6 overestimate the loads. Thus, a β of 0.5 is proper for aluminum 6061-T6. According to experimental results of mild steel (Jia and Kuwamura 2014b), β can be taken as a constant of 0.5, though it has a minor dependence on strain range. In practice, a β of 0.5 is recommended if no cyclic test results are available, and this parameter can also be calibrated through cyclic test results at different strain ranges. Besides,

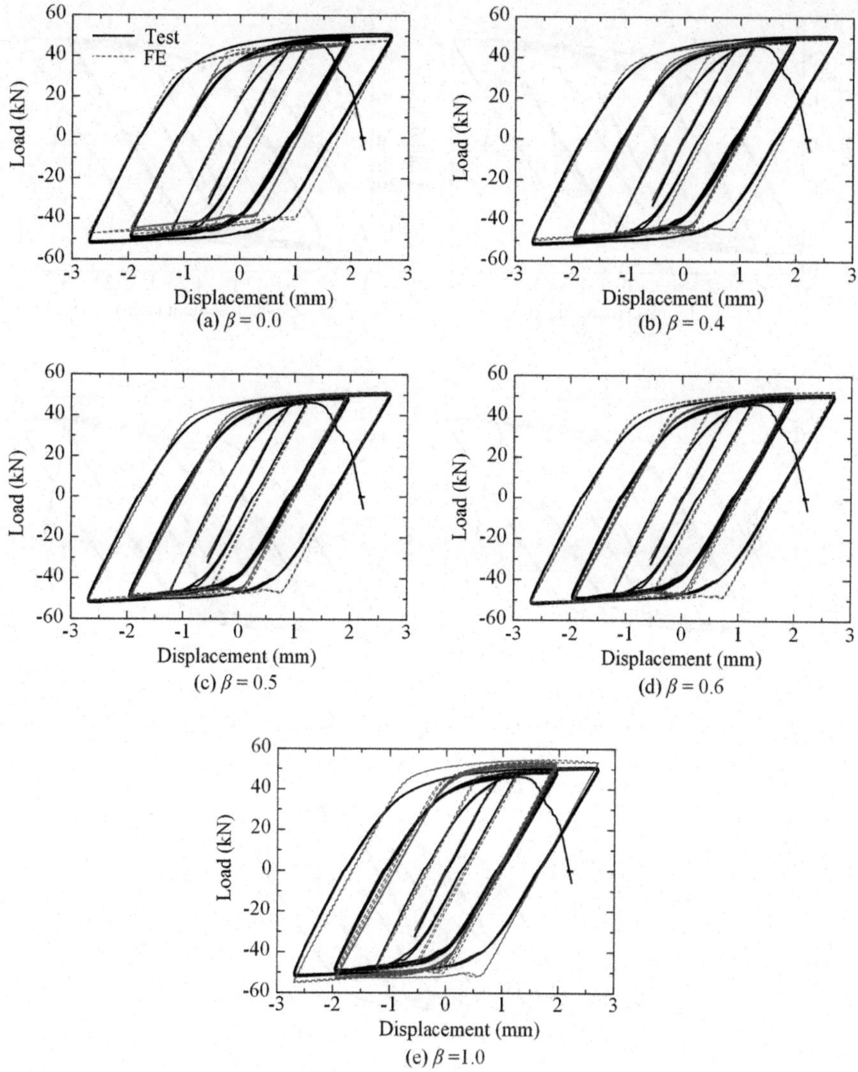

Fig. 9.16 Comparison of load–displacement curves for specimen S2-L4S20G1

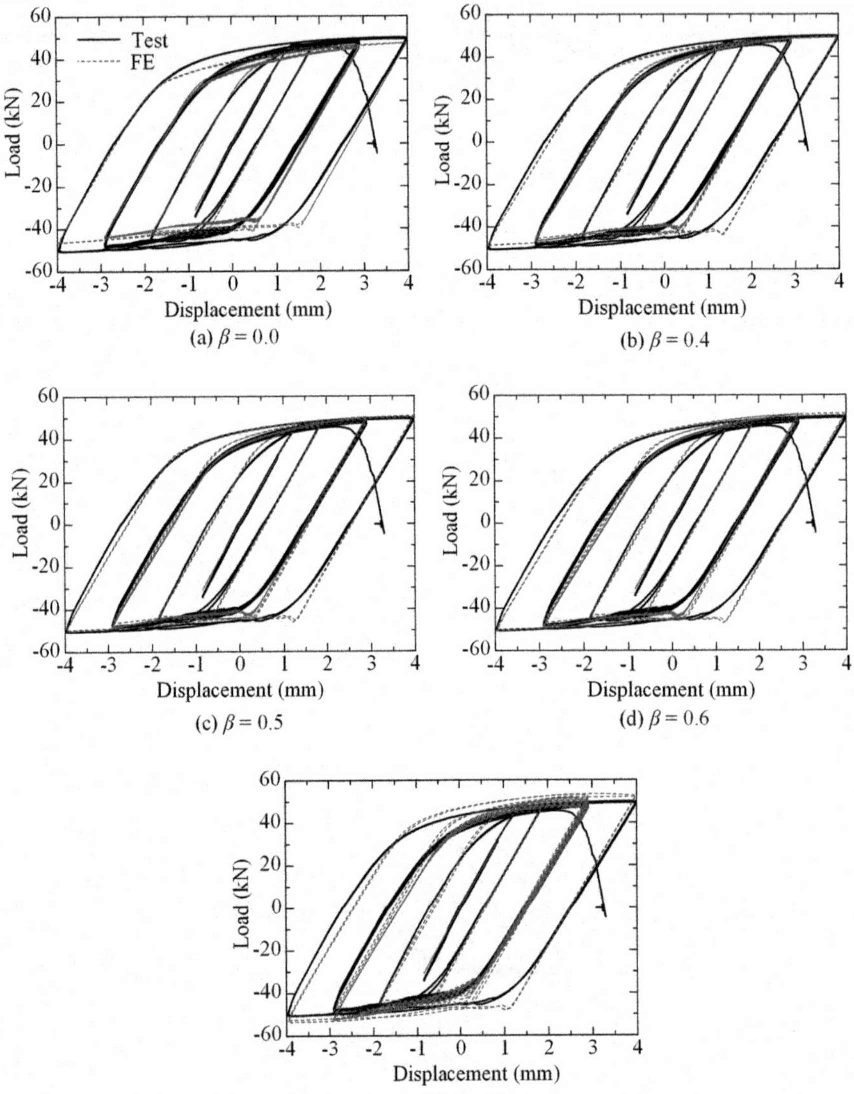

Fig. 9.17 Comparison of load–displacement curves for specimen S2-L6S20G1

the stresses at the elastic–plastic transition regions are still overestimated except for the first loading half cycle. This is also mainly due to the fact that the hardening rate at the elastic–plastic transition region under monotonic tension is larger than that of the subsequent cyclic loading half cycles.

9.5 Summaries

Accuracy of a cyclic plasticity model is determined by not only the formation of the model itself, but also the corresponding calibration method. This chapter presents a new method to calibrate the widely used generalized Armstrong–Frederick model with an isotropic hardening (IH) rule implemented, which is also termed as the Chaboche model with IH in this study. The newly proposed method was validated at both material and member levels, respectively, using experimental results of double-edge-notched (DEN) specimens and aluminum buckling-restrained braces (BRBs).

(1) For structural aluminum 6061-T6, the plasticity model with two backstresses can well describe the cyclic behaviors under cyclic large plastic strain loading at the full strain range, where a linear formation is employed for one of the backstresses.
(2) Plastic reversals with the same displacement amplitude at the post-necking stage tend to cause larger damage to the aluminum compared with those at the pre-necking stage. As the number of reversals increases, the rupture displacement can also be greatly reduced due to the cumulative damage.
(3) The plasticity model with the newly proposed calibration method can well simulate plasticity of structural aluminum under variable-amplitude cyclic loading at the full strain range.
(4) The plasticity model with the newly proposed calibration method can well evaluate the buckling modes and load–displacement curves of aluminum BRBs under cyclic large plastic strain loading, where the plasticity model parameters can be calibrated using only the yield strength and the tensile strength. A β of about 0.5 gives the best evaluation result.
(5) Deviation at the elastic–plastic transition regions was observed, which is mainly due to the different hardening rates of the first tensile half cycle and the subsequent loading half cycles.
(6) For aluminum 6061-T6 at the material level, the proportion of the IH component in the total hardening stress, β, with a value of about 0.4 gives the most accurate evaluation results under small plastic strain ranges.
(7) For aluminum 6061-T6 at the material level, a β of about 0.5 gives the best evaluation result at the tension side, while a β of about 0.6 for the compression side under medium and large plastic strain ranges, e.g., strain close to and beyond the necking initiation strain.

References

ABAQUS (2010) ABAQUS standard manual (version 6.10). Karlsson & Sorensen Inc., Hibbitt, Pawtucket

Brando G et al (2013) Experimental tests of a new hysteretic damper made of buckling inhibited shear panels. Mater Struct 46:2121–2133

Chang KC, Lee GC (1986a) Biaxial properties of structural steel under nonproportional loading. J Eng Mech (ASCE) 112

Chang KC, Lee GC (1986b) Constitutive relations of structure steel under nonproportional loading. J Eng Mech (ASCE) 112

Dørum C et al (2010) Finite element analysis of plastic failure in heat-affected zone of welded aluminium connections. Comput Struct 88:519–528

De Matteis G et al (2000) T-stub aluminium joints: influence of behavioural parameters. Comput Struct 78:311–327

De Matteis G et al (2008) Numerical and experimental analysis of pure aluminium shear panels with welded stiffeners. Comput Struct 86:545–555

Dey P et al (2016) Evaluation of design guidelines for the serviceability assessment of aluminum pedestrian bridges. J Bridge Eng (ASCE) 22:04016109

Frederick CO, Armstrong PJ (2007) A mathematical representation of the multiaxial Bauschinger effect. Mater High Temp 24:1–26

Ge H, Kang L (2014) Ductile crack initiation and propagation in steel bridge piers subjected to random cyclic loading. Eng Struct 59:809–820

Hu F et al (2016a) Constitutive model for full-range elasto-plastic behavior of structural steels with yield plateau: calibration and validation. Eng Struct 118:210–227

Hu F et al (2016b) Constitutive model for full-range elasto-plastic behavior of structural steels with yield plateau: Formulation and implementation. Eng Struct (In press)

Jia L-J, Kuwamura H (2014a) Ductile fracture simulation of structural steels under monotonic tension. J Struct Eng (ASCE) 140:04013115

Jia L-J, Kuwamura H (2014b) Prediction of cyclic behaviors of mild steel at large plastic strain using coupon test results. J Struct Eng (ASCE) 140:04013056

Jia L-J, Kuwamura H (2015) Ductile fracture model for structural steel under cyclic large strain loading. J Constr Steel Res 106:110–121

Jia L-J et al (2016a) Experimental and numerical study on ductile fracture of structural steels under combined shear and tension. J Bridge Eng (ASCE):04016008

Jia L-J et al (2016b) Ductile crack initiation and propagation of structural steels under cyclic combined shear and normal stress loading. Constr Build Mater 112:69–83

Kang L et al (2015) Experimental and ductile fracture model study of single-groove welded joints under monotonic loading. Eng Struct 85:36–51

Khadyko M et al (2015) Simulation of large-strain behaviour of aluminium alloy under tensile loading using anisotropic plasticity models. Comput Struct 157:60–75

Kuhlmann-Wilsdorf D, Laird C (1979) Dislocation behavior in fatigue II. Friction stress and back stress as inferred from an analysis of hysteresis loops. Mat Sci Eng 37:111–120

Liao F et al (2015) Ductile fracture prediction for welded steel connections under monotonic loading based on micromechanical fracture criteria. Eng Struct 94:16–28

Liu Y et al (2017) Ductile-fatigue transition fracture mode of welded T-joints under quasi-static cyclic large plastic strain loading. Eng Fract Mech 176:38–60

Matteis GD et al (2001) Cross-sectional classification for aluminum beams—parametric study. J Struct Eng (ASCE) 127:271–279

Mazzolani FM et al (2011) Local buckling of aluminum alloy angles under uniform compression. J Struct Eng (ASCE) 137:173–184

Moen LA et al (1999) Rotational capacity of aluminum beams under moment gradient. II: numerical simulations. J Struct Eng (ASCE) 125:921–929

Rosien FJ, Ostertag CP (2009) Low cycle fatigue behavior of constraint connections. Mater Struct 42:171–182
Saleem MA et al (2012) Experimental evaluation of aluminum bridge deck system. J Bridge Eng (ASCE) 17:97–106
Shen C et al (1995) Cyclic behavior of structural steels. II: theory. J Eng Mech (ASCE) 121
Su M-N et al (2015) Continuous beams of aluminum alloy tubular cross sections. I: tests and FE model validation. J Struct Eng (ASCE) 141:04014232
Su M-N et al (2016) The continuous strength method for the design of aluminium alloy structural elements. Eng Struct 122:338–348
Tabatabai H, Hawileh R, Rahman A (2010) Evaluation of the low-cycle fatigue life in ASTM A706 and A615 grade 60 steel reinforcing bars. J Mater Civ Eng 22
Ucak A, Tsopelas P (2011) Constitutive model for cyclic response of structural steels with yield plateau. J Struct Eng (ASCE) 137
Ucak A, Tsopelas P (2012) Accurate modeling of the cyclic response of structural components constructed of steel with yield plateau. Eng Struct 35:272–280
Wang J et al (2016) Constitutive model of low-yield point steel and its application in numerical simulation of buckling-restrained braces. J Mater Civ Eng 28
Wang CL et al (2018) Concept and performance testing of an aluminum alloy bamboo-shaped energy dissipater. Struct Des Tall Spec Buildings 27:e1444
Yin S et al (2004) Degradation and buckling of I-beams under cyclic pure bending. J Eng Mech (ASCE) 130:809–817
Yoshida F, Uemori T (2002) A model of large-strain cyclic plasticity describing the Bauschinger effect and workhardening stagnation. Int J Plast 18:661–686

Chapter 10
Ultra-low-Cycle Fatigue Failure of Aluminum

Abstract Ultra-low-cycle fatigue (ULCF) life of ductile metal is closely correlated with monotonic tension coupon test results, since the dominant failure mode is the same. In this chapter, a novel approach to evaluate crack initiation of aluminum alloy under ULCF loading only using monotonic tension coupon test results is proposed. ULCF tests on 15 specimens made of aluminum alloy 6061-T6 are conducted, and numerical analyses using the previously proposed cyclic ductile fracture model indicate that the ULCF life of aluminum can be greatly underestimated. A new fracture model based on the concept of different dislocation structures is thus proposed, which classifies damage into kinematic hardening correlated and isotropic hardening correlated. A material constant is employed to consider the relatively low damage induced by the kinematic hardening compared with the isotropic hardening one. The newly proposed fracture model can well simulate the instants of crack initiation for the specimens. A process to evaluate the ULCF life of aluminum alloy based on both tension coupon test results and simple numerical analysis is presented.

10.1 Introduction

Studies on high- and low-cycle fatigue lives of a variety of types of aluminum alloy (González et al. 2011; Hao et al. 2014; May et al. 2013; Naeimi et al. 2017) and welded joints (Al Zamzami and Susmel 2017; Ambriz et al. 2010) have been extensively conducted. It has been found that the fracture mode of ductile metal can transit from fatigue fracture to ductile fracture as the strain amplitude increases to a critical value (Kuwamura 1997). There is also a transition mode with mixed fatigue fracture and ductile fracture mode for the case when the strain amplitude is medium compared with those of fatigue fracture and ductile fracture (Panontin and Sheppard 1995).

For aluminum building structures as shown in Fig. 10.1 in regions with a high seismic risk, the structures can fail after sustaining dozens or hundreds of large plastic loading reversals, which is termed as ultra-low-cycle fatigue (ULCF) fracture in this study. Under this circumstance, the failure mode is either ductile fracture or transition fracture mode between ductile fracture and fatigue fracture. A number of

L.-J. Jia and H. Ge, *Ultra-low-Cycle Fatigue Failure of Metal Structures under Strong Earthquakes*, Springer Tracts in Civil Engineering,
https://doi.org/10.1007/978-981-13-2661-5_10

studies were conducted on ductile fracture of metals and specimens using ductile fracture models (e.g., Bai and Wierzbicki 2015; Bao and Wierzbicki 2005; Jia et al. 2014, 2016a, b; Jia and Kuwamura 2014; Khandelwal and El-Tawil 2014; Kiran and Khandelwal 2013, 2014; Xue and Wierzbicki 2008), and a cyclic void growth model (CVGM) (Jia and Kuwamura 2015) was also employed to investigate ductile fracture of steel specimens under cyclic incremental-amplitude loading. It was also found that there is also a correlation between high-cycle fatigue strength and tension coupon test results for metals (Özdeş and Tiryakioğlu 2017; Pang et al. 2013). These research achievements are of great importance from the viewpoint of engineers since commonly engineers can only obtain tension coupon test results in practice. However, to date, there is still no research on correlation between tension coupon test results and ULCF, to the authors' knowledge.

This study aims to bridge the gap between monotonic tension coupon test results and evaluation of ULCF based on test results of aluminum specimens. In this chapter, a novel damage model is proposed by distinguishing the damage into kinematic hardening-related and isotropic hardening-related components, where the damage accumulation rule still follows the CVGM. Numerical analysis using the CVGM is conducted, the results of which indicate that the CVGM greatly overestimates the number of loading cycles till crack initiation. The newly proposed damage model can well evaluate the instant of crack initiation, and corresponding model parameters for aluminum alloy 6061-T6 are also determined.

10.2 Ultra-low-Cycle Fatigue Tests on Aluminum Alloy 6061-T6

10.2.1 Experimental Program

A constant-amplitude cyclic loading protocol shown in Fig. 10.2 was devised. Experimental tests on 15 double-edge-notched (DEN) specimens the same as that illustrated

(a) Nanjing Usnisa Palace (b) Tianjin Bengbu Bridge

Fig. 10.1 Applications of aluminum in spatial structures and bridges

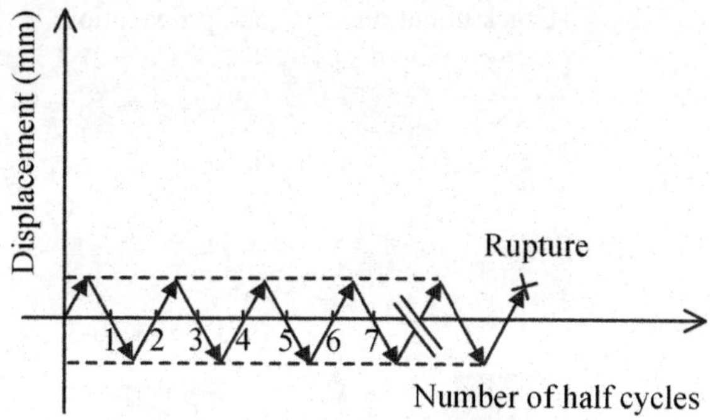

Fig. 10.2 Loading history for double-edge-notched specimens

in Fig. 9.3 were conducted under cyclic loading till rupture. Considering instability of the specimens under large compressive strain loading, the specimens were designed as compact as possible to avoid premature elasto-plastic buckling. The cross section of the central part of the DEN specimen is uniform with a length of 5 mm to make the strain distribution as uniform as possible, though it is still too short to achieve the aim due to the instability problem. The width of the central cross section is 20 mm and that of the end is 40 mm to attach a specially designed extensometer.

Three coupon tests as illustrated in Fig. 9.4 were also carried out to obtain the mechanical properties of the material. All the specimens and the coupons were manufactured from the same 13.5-mm-thick extruded aluminum alloy 6061-T6 plate. The average mechanical properties obtained from the three coupon tests and chemical composition of the material is the same as that listed in Table 9.1. The average yield strength is 257 MPa, and the tensile strength is 300 MPa, which indicates a yield-to-tensile strength ratio of 0.86. The gage length of the coupon test is 50 mm, and the average elongation is 12%, which is around half of that of mild steel. The test setup of the DEN specimens is shown in Fig. 9.5. The bottom end of each specimen was fixed, and the longitudinal freedom of the top end was movable. The rotational freedom of the top end along the longitudinal direction was relaxed, while the other two rotational freedoms were not. All the tests were conducted using an MTS loading device as shown in Fig. 9.5 at a quasi-static speed at room temperature around 30°. All the DEN specimens were under cyclic constant-amplitude loading, and the tests were all automatically controlled by the displacement data of an MTS extensometer with a gage length of 50 mm as shown in Fig. 9.5. The loading history is illustrated in Fig. 10.2, and the expected and actual displacement amplitudes were also given in Table 10.1. The tests were automatically controlled by the loading system, and it can be found that the maximum deviation between the expected displacements and the actual ones is within 3.5%. Five different strain amplitudes, i.e., 3.10, 3.45, 3.80, 4.15, and 4.50 were designed. These strain amplitudes were determined based on preliminary finite element analyses using the previously proposed CVGM by one of the authors (Jia and Kuwamura 2015). The positive and negative peak strains, respectively, correspond to the maximum equivalent strain of the specimen under tension and compression. The main target of the preliminary analyses was to ensure

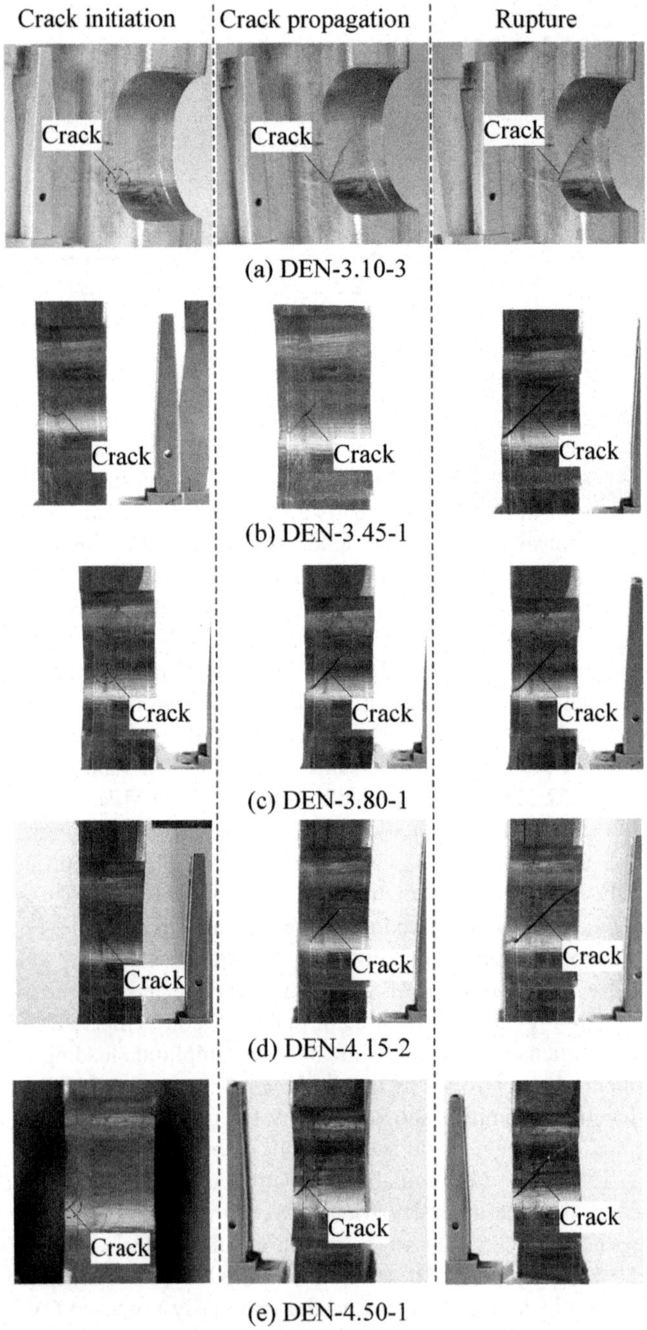

Fig. 10.3 Failure process of DEN specimens

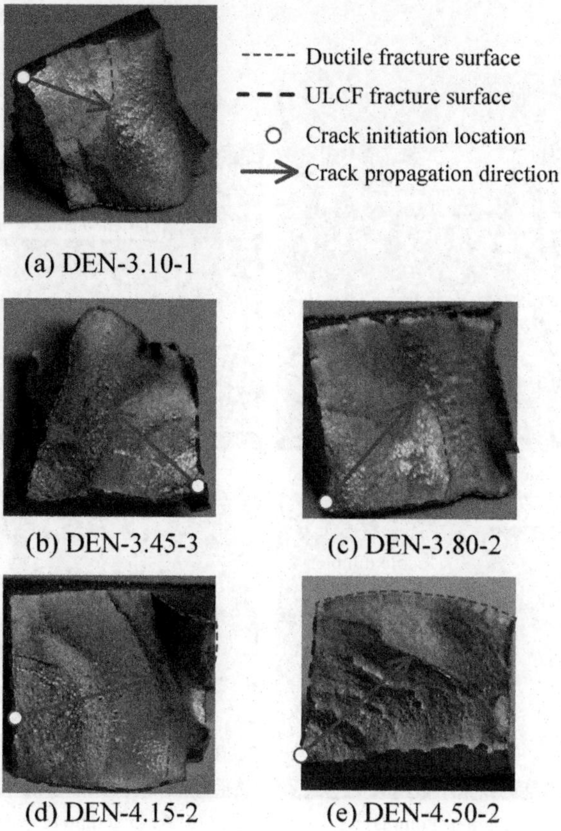

Fig. 10.4 Macro-fracture surfaces of DEN specimens

Table 10.1 Expected displacement amplitudes of different specimens (*Unit* mm)

Specimens	DEN-3.10	DEN-3.45	DEN-3.80	DEN-4.15	DEN-4.50
Expected displacement amplitude	±0.378	±0.403	±0.429	±0.456	±0.483
Actual displacement amplitude	+0.367	+0.395	+0.420	+0.445	+0.469
	−0.367	−0.393	−0.419	−0.444	−0.466

the specimens fail within a fatigue life of several to dozens of loading cycles. For each strain amplitude, three specimens were tested to investigate the deviation due to possible geometrical, mechanical imperfections, and other possible factors. The naming of the specimens is illustrated using specimen DEN-3.10-2, where "3.10" denotes that the equivalent strain amplitude of the point with the maximum equivalent strain is 3.10%, and "2" denotes the second specimen of the DEN-3.10 series.

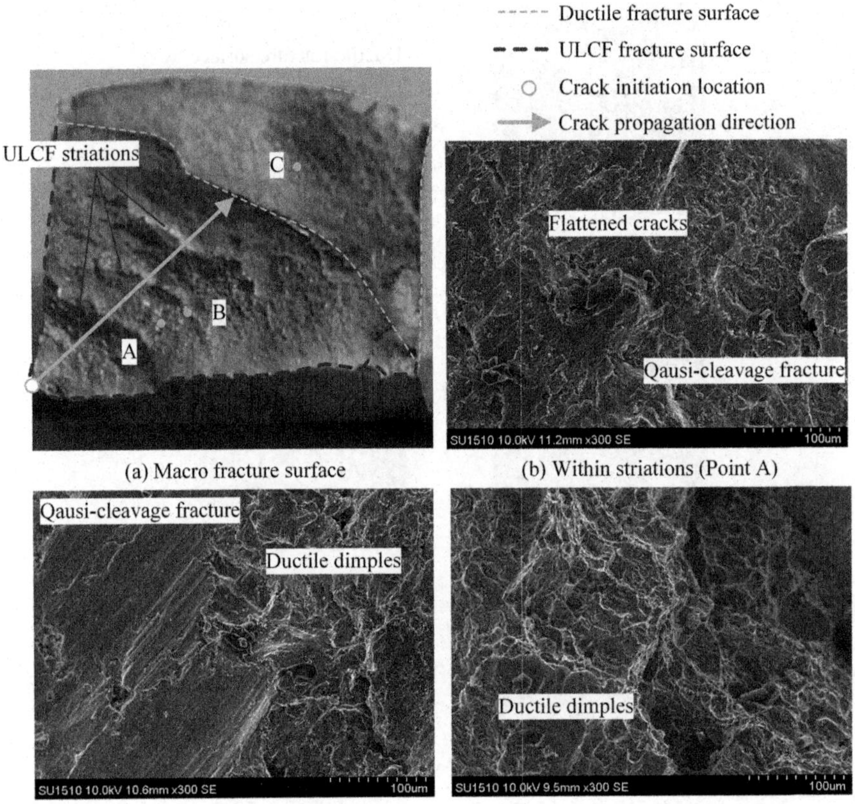

(a) Macro fracture surface (b) Within striations (Point A)

(c) On striation (Point B) (d) At final rupture zone (Point C)

Fig. 10.5 Fractographic study of a specimen (DEN-4.50-2)

10.2.2 Failure Modes and Fractographic Study

During the experiments, the testing was temporarily paused at both tensile and compressive peak displacements to verify the cracking states of the specimens. Failure of all the specimens is characterized by cracking, where the cracking processes of typical specimens are illustrated in Fig. 10.3. Failure of all the specimens started from crack initiation at the surface as shown in the figure, where the crack initiation was defined when the crack can be visualized by human eyes. The crack then propagated along a diagonal direction and developed into the width direction. All the specimens failed due to rupture of the whole cross sections when the main cracks developed to a substantial size. Macro-fracture surfaces observed using a conventional camera are given in Fig. 10.4, indicating non-conventional ULCF fracture surfaces. Fracture surfaces of high- and low-cycle fatigue bodies are characterized by a large number of fatigue striations. The ones of ULCF problems are distinguished from the conventional ones, where several striations with substantial widths can be observed as shown in Fig. 10.4. The crack initiation locations can also be deduced from the macro-fracture surfaces as shown in Fig. 10.4. All the cracks initiated from the surfaces of the specimens. The macro-fracture surfaces can be generally divided into two different zones, i.e., ULCF fracture surface and ductile fracture surface, as shown in Fig. 10.4. The former is represented by concentric arcs and the latter by dimples. The ductile fracture surface is formed due to final rupture of the DEN specimens when the crack increases to a certain size. Similar failure modes were observed in ULCF test results of mild steel (Liu et al. 2017), where several wide striations can be found, and ductile dimples can be observed on the striations using a scanning electron microscope (SEM). This indicates that the ULCF failure mechanism is mainly dominated by ductile fracture. This implies the possibility to utilize monotonic tension coupon test results to evaluate ULCF life of ductile metal.

To verify the fracture modes from the micro-level, fractographic study using an SEM was conducted for the ULCF fracture surfaces. The observation results of the three typical locations on the fracture surface of specimen DEN-4.50-2 as shown in Fig. 10.5a are given in Fig. 10.5b–d. The three typical locations are, respectively, within two ULCF striations, on a ULCF striation, and on the ductile fracture surface. For Point A located within the striations, there are flat and smooth surfaces indicating quasi-cleavage fracture as shown in Fig. 10.5c, and a number of tiny dimples can also be found in Fig. 10.5b. The tiny dimples are flattened under the cyclic loading. The fracture surface indicates a quasi-brittle fracture mode with both tiny flattened ductile dimples and brittle quasi-cleavage fracture. The fracture surface of Point B as shown in Fig. 10.5c consists of two distinguished zones, where the left one is similar to that of Fig. 10.5b and the right one indicates a ductile dimple pattern. It is deduced that the ductile dimples are formed due to large strain amplitude during the crack propagation of the DEN specimen. The Point C has a ductile dimple pattern as shown in Fig. 10.5d, where the typical dimple size is 10–100 μm, which is much less than that of mild steel.

10.2.3 Hysteretic and Skeleton Curves

The load–displacement curves and the corresponding skeleton curves of the DEN specimens are, respectively, given in Figs. 10.6 and 10.7. Figure 10.6 implies stable hysteretic properties of all the DEN specimens. The instants of crack initiation are also marked on the skeleton curves, indicating that cracks all initiated from compressive loading half cycles. The skeleton curves shown in Fig. 10.7 indicate that the load-carrying capacity first increases owing to strain hardening during the first several loading cycles and then becomes stabilized before cracking. After crack initiation, the tensile load-carrying capacities decrease gradually, while the compressive ones are almost unaffected. The whole failure process indicates a ductile manner. The peak tensile and compressive loads with their average values are given in Table 10.2, where the average peak tensile loads range from 84.2 to 88.7 kN and the compressive ones from −85.6 to −92.1 kN. The numbers of half cycles till crack initiation and

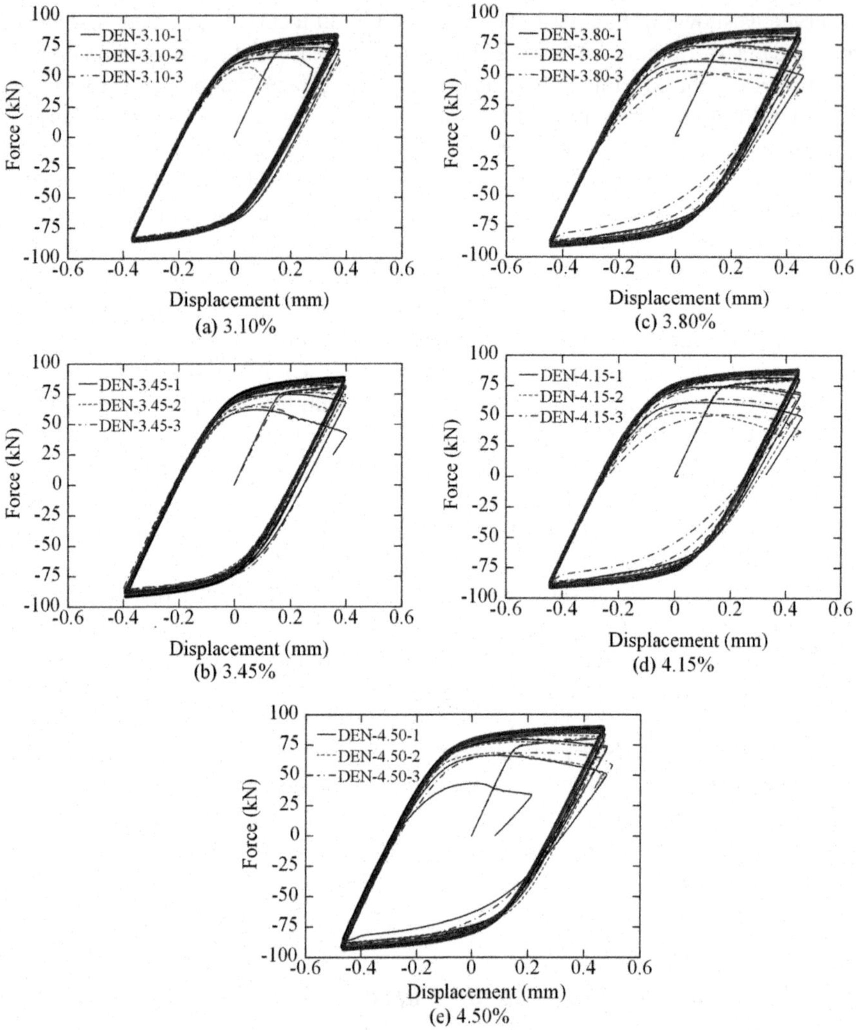

Fig. 10.6 Load–displacement curves of DEN specimens

final failure are also listed in the table. In this chapter, the final failure is defined when the load decreases to less than 85% of the peak load. The average number of half cycles till crack initiation, ${}_tN_{\text{ini, ave}}$, decreases from 87 to 49 and that till final failure, ${}_tN_{f,\text{ ave}}$, decreases from 106 to 58 when the strain amplitude increases from 3.10 to 4.50% as shown in Table 10.2. The crack propagation stage takes a significant portion of the ULCF life, ranging from 15.5 to 22.9% for the five strain amplitudes.

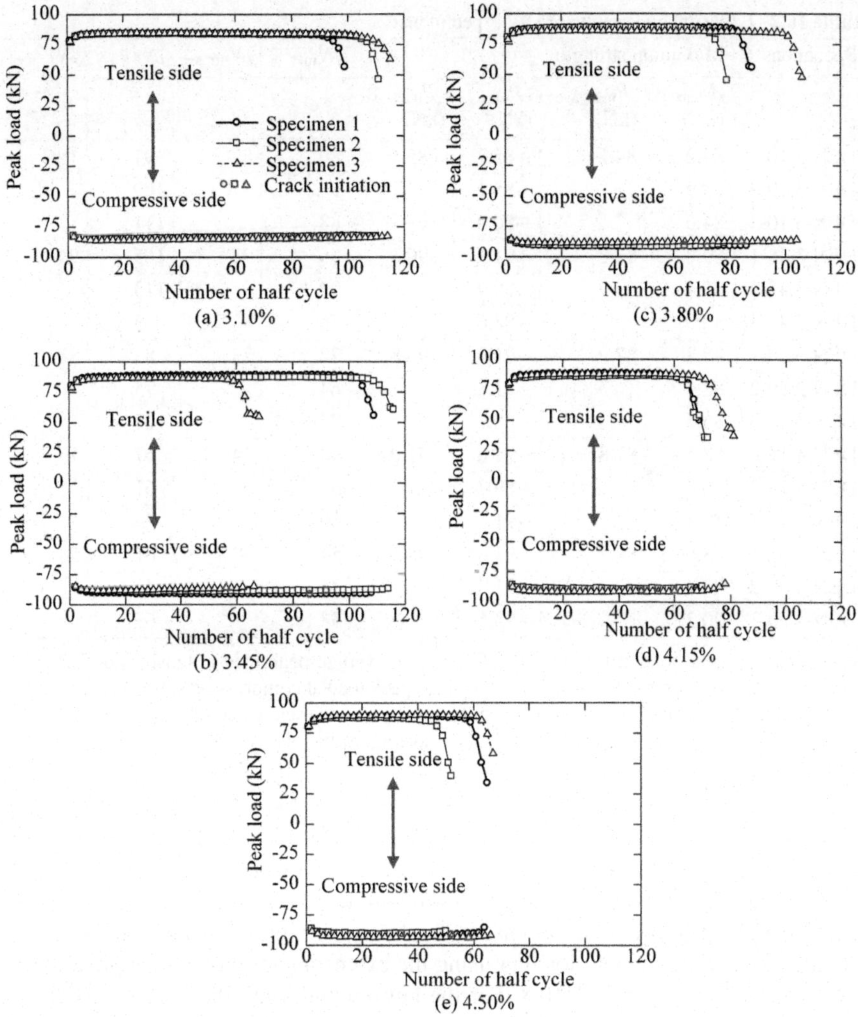

Fig. 10.7 Peak load versus number of half cycles for specimens

10.3 Numerical Simulation

10.3.1 FE Modeling

Based on the experimental results, it can be found that the main failure mechanism is ductile fracture for the ULCF problems. Thus, the CVGM in Chap. 6 was employed to evaluate crack initiation of the DEN specimens. The finite element software ABAQUS was employed to carry out the numerical simulation. Three-

Table 10.2 Experimental results of DEN specimens

Specimens	Maximum strength				${}_tN_{\text{ini}}$	${}_tN_{\text{ini, ave}}$	${}_tN_f$	${}_tN_{f,\text{ ave}}$
	${}_tP_{\max}$ (kN)	${}_tP_{\max,\text{ ave}}$ (kN)	${}_cP_{\max}$ (kN)	${}_cP_{\max,\text{ ave}}$ (kN)				
DEN-3.10-1	84.2	84.2	−85.7	−85.6	80	87	97	106
DEN-3.10-2	83.9		−85.1		94		109	
DEN-3.10-3	84.6		−86.0		88		111	
DEN-3.45-1	89.0	87.8	−91.9	−90.1	86	78	105	94
DEN-3.45-2	87.4		−89.9		92		113	
DEN-3.45-3	87.0		−88.6		56		63	
DEN-3.80-1	88.8	87.9	−91.9	−90.8	72	73	85	88
DEN-3.80-2	88.6		−91.8		64		75	
DEN-3.80-3	86.3		−88.7		84		103	
DEN-4.15-1	88.5	87.8	−91.6	−91.0	48	54	67	70
DEN-4.15-2	86.6		−89.7		54		67	
DEN-4.15-3	88.4		−91.6		60		75	
DEN-4.50-1	88.6	88.7	−91.8	−92.1	54	49	61	58
DEN-4.50-2	87.4		−91.0		34		49	
DEN-4.50-3	90.2		−93.5		58		65	

Notes ${}_tP_{\max}$ = peak load at tensile side; ${}_tP_{\max,\text{ ave}}$ = average peak load at tensile side; ${}_cP_{\max}$ = peak load at compressive side; ${}_cP_{\max,\text{ ave}}$ = average peak load at compressive side; ${}_tN_{\text{ini}}$ = number of half cycles till crack initiation of test results; ${}_tN_{\text{ini, ave}}$ = average value of ${}_tN_{\text{ini}}$; ${}_tN_f$ = number of half cycles till failure of test results; ${}_tN_{f,\text{ ave}}$ = average value of ${}_tN_f$

dimensional solid models as illustrated in Fig. 10.8 were established, and the element type was selected as C3D8R considering efficiency and accuracy. The portion of DEN specimen within the gage length was simulated, and the boundary conditions were also given in the figure to ensure similar constraining effects as those of the experiments. Displacement loading using the extensometer data was applied to the specimens as shown in Fig. 10.8. For the bottom end, only the axial translational freedom was constrained, and the central node of the cross section was fixed to avoid rigid rotation of the whole model.

For the fracture parameter χ_{cr}, the value can be calibrated through the corresponding tension coupon test results, which is 0.9 as listed in Table 10.3. For the cyclic plasticity model, the Chaboche model with IH was employed. The calibrated material constants for the plasticity model using the corresponding coupon test results are given in Table 10.3.

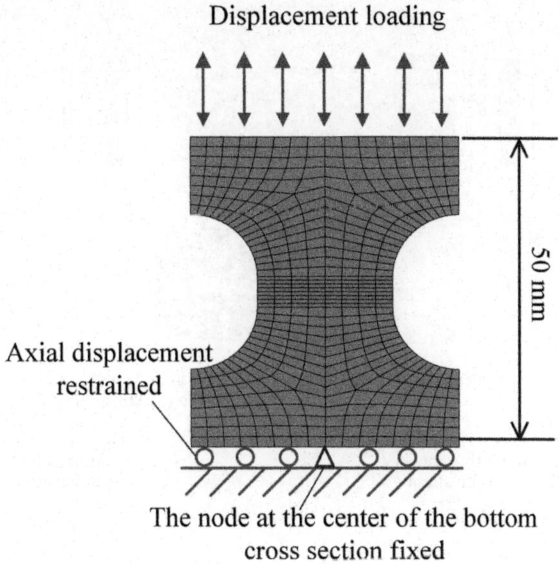

Fig. 10.8 FE model and boundary conditions for double-edge-notched specimens

10.3.2 Numerical Results and Comparison with Experimental Results

Due to the compact configuration of the DEN specimens, the stress and strain distributions are not uniform at the central cross section. The contour plots of the stress triaxiality, equivalent plastic strain, and D_{CVGM} at the peak tensile and compressive deformations of the first cycle are given in Fig. 10.9. Figure 10.9a indicates that the stress triaxiality concentrates at the center of the cross section. Figure 10.9b shows that the equivalent plastic strain distribution is more uniform compared with those of the stress triaxiality. It can be found from Fig. 10.9c that damage concentrates at the central part of the cross section at the end of the tensile half cycle, while the damage distribution becomes more uniform across the width at the end of the compressive half cycle. This explains why the cracks can initiate from the surface of the DEN specimens during the experiments.

Stress triaxiality history curves of two critical points on the central cross section are plotted in Fig. 10.10. Points A and B are, respectively, located at the center and surface of the cross section. It can be found that the stress triaxiality at the center (Point A) can increase as much as about 6.0, which is an extremely high value

Table 10.3 Calibrated model parameters of plasticity model

Specimen	β	C_1 (MPa)	Q_∞ (MPa)	σ_{y0} (MPa)	C_2 (MPa)	γ_1	k	γ_2	χ_{cr}
DEN	0.5	750	48	257	25	18.5	14.1	0	0.9

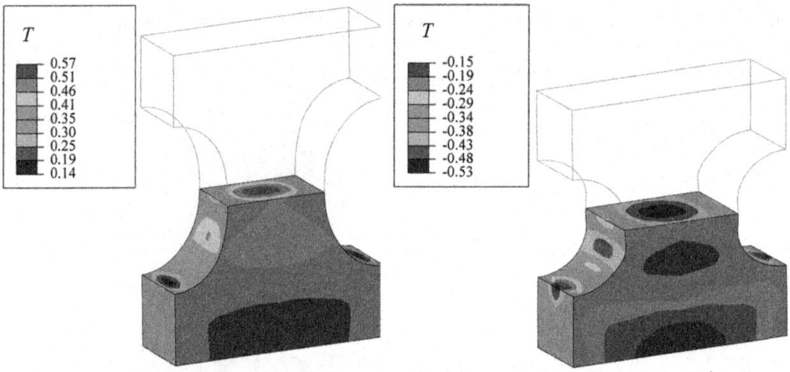

Peak tensile deformation of 1st cycle Peak compressive deformation of 1st cycle

(a) Stress triaxiality

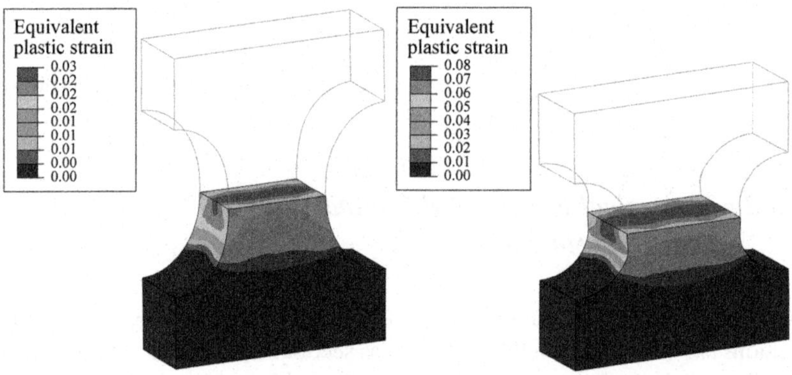

Peak tensile deformation of 1st cycle Peak compressive deformation of 1st cycle

(b) Equivalent plastic strain

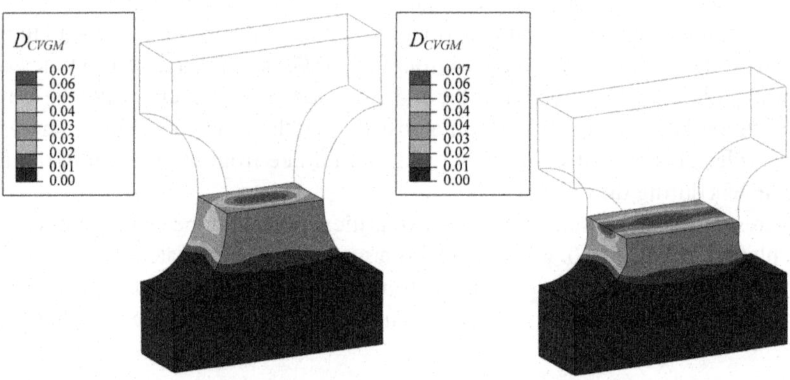

Peak tensile deformation of 1st cycle Peak compressive deformation of 1st cycle

(c) Damage index D_{CVGM}

Fig. 10.9 Contour plots of specimens

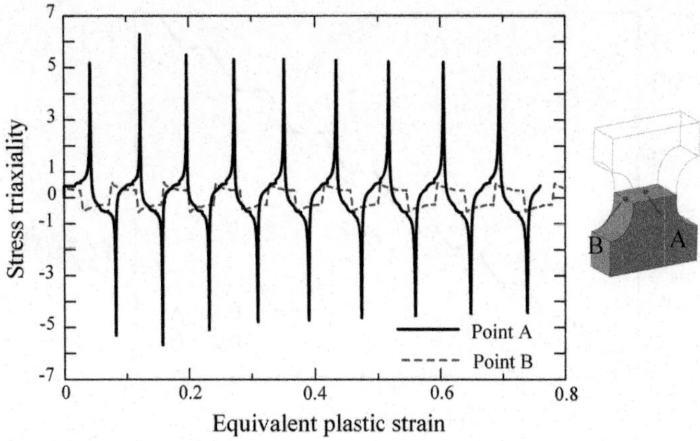

Fig. 10.10 Equivalent plastic strain versus stress triaxiality curves

corresponding to very small fracture strain. For the one on the surface (Point B), the stress triaxiality is less than 1.0, which is much lower than that at the center.

The accumulation histories of D_{CVGM} of the two critical points are also given in Fig. 10.11. For the central Point A, D_{CVGM} stops increasing at compressive loading half cycles as shown in Figs. 10.9c and 10.11, since the stress triaxiality at the center is below $-1/3$, and no damage accumulates according to the CVGM. For the surface Point B shown in Fig. 10.10, D_{CVGM} still increases during the compressive loading half cycles, and this explains why cracks can initiate from the surface of the specimens in the experiments. The difference is mainly due to the different stress triaxiality states. For the surface Point B, the stress triaxiality changes from negative to positive during the compressive loading half cycle as shown in Fig. 10.10, and D_{CVGM} thus can go on increasing during the compressive loading half cycles. In addition, Fig. 10.10 also indicates that the equivalent plastic strain−stress triaxiality curve at the third loading cycle is almost the same as those of the subsequent loading cycles. Based on this observation, one can just employ the analysis results of the third loading cycle to evaluate the corresponding results of a specimen during the whole loading history if the loading protocol is cyclic constant-amplitude.

The numbers of half cycles till crack initiation for all the DEN specimens predicted by the CVGM, ${}_{t}N_{\mathrm{ini,\,CVGM}}$, are listed in Table 10.4, which are compared with the corresponding experimental ones. The comparison results indicate that the CVGM greatly underestimates the numbers of half cycles till crack initiation. The previously proposed CVGM has just been validated by experimental results with a failure life of several loading cycles, and the loading protocols are mainly with incremental-amplitude. The current comparison results show that the CVGM cannot well evaluate the cases under constant-amplitude loading with a failure life at a level of dozens of loading cycles.

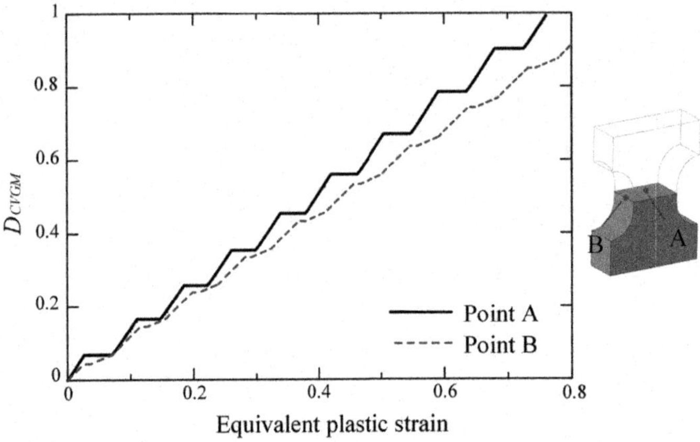

Fig. 10.11 Equivalent plastic strain versus damage index of CVGM curves

10.4 Proposed New Method to Evaluate Fatigue Strength at Ultra-low-Cycle Fatigue Regime

It is known that damage is closely correlated with plastic deformation, especially for ductile metal. Strain hardening of common ductile metal can be generally divided into two types, i.e., IH and KH. IH is commonly induced by the formation of stable dislocation structures such as cell walls, while KH is mainly triggered by dislocations with less stable structures such as piled-up dislocations (Yoshida and Uemori 2002). The definitions of IH and KH are illustrated in Fig. 10.12 using a typical cyclic stress–strain curve of ductile metal under uniaxial loading. Figure 10.12a shows a negligible stress hardening effect during the fifth loading reversal, while significant strain hardening occurs during the third loading reversal. The main reason is that the

Table 10.4 Comparison of number of half cycles till crack initiation

Specimens	${}_tN_{\text{ini, ave}}$	${}_tN_{\text{ini, theor}}$	${}_tN_{\text{ini, CVGM}}$	${}_tN_{\text{ini, theor}}/{}_tN_{\text{ini, ave}}$	${}_tN_{\text{ini, CVGM}}/{}_tN_{\text{ini, ave}}$
DEN-3.10	87	84	31	0.97	0.36
DEN-3.45	78	73	27	0.94	0.35
DEN-3.80	73	68	23	0.93	0.32
DEN-4.15	54	60	21	1.11	0.39
DEN-4.50	49	55	19	1.12	0.39
			Average	1.01	0.36
			CoV	0.094	0.086

Notes ${}_tN_{\text{ini, ave}}$ = average number of half cycles till crack initiation of test results; ${}_tN_{\text{ini, theor}}$ = expected number of half cycles till crack initiation according to the newly proposed theoretical equation; ${}_tN_{\text{ini, VGM}}$ = number of half cycles till crack initiation according to the cyclic void growth ductile fracture model

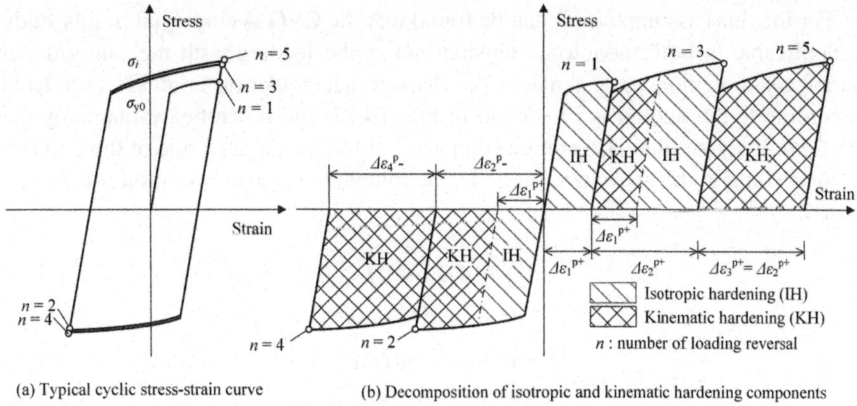

Fig. 10.12 Decomposition of isotropic and kinematic hardening correlated damage

strain amplitude doubles at the third loading reversal compared with the first one, while the strain amplitude remains unchanged at the fifth one compared with the third one. Strain hardening is closely correlated with damage of material. To consider the different damage accumulation rates, the IH correlated damage is defined when there is an increase in the equivalent plastic strain amplitude compared with previous loading reversals, and the other portion is defined as the KH correlated one as shown in Fig. 10.12.

The number of loading half cycles till crack initiation is termed as crack initiation life in this study. The CVGM greatly underestimates the crack initiation life under cyclic constant-amplitude loading, and it is known that KH is dominant for the case of cyclic constant-amplitude loading. This implies that the damage correlated with KH is overestimated. Thus, an assumption is proposed in this study.

(1) For the same equivalent plastic strain increment with the same stress triaxiality, damage accumulation due to IH is larger than that of KH.
(2) Damage accumulation due to KH is proportional to that of IH.
(3) IH correlated damage follows the accumulation rule of the CVGM.

The first assumption can be easily accepted, since it has been observed in a number of cyclic test results of metal (e.g., Kuroda 2002; Ohata and Toyoda 2004). The experimental and numerical results in this study also prove rationality of this assumption.

For the second assumption, it is postulated that no damage accumulates for KH correlated plastic straining in the literature (Ohata and Toyoda 2004), which is not applicable to the cases under cyclic constant-amplitude loading. According to the damage rule in a previous study (Ohata and Toyoda 2004), ductile metal will never rupture under cyclic constant-amplitude loading, which is not the fact based on experimental findings. As a first approximation, damage accumulation due to KH is postulated to be linear with that of IH.

For the third assumption, it can be found that the CVGM employed in this study is applicable to both monotonic tension and cyclic loading with the same model parameter. For monotonic tension, the damage accumulation is all IH correlated according to the definition illustrated in Fig. 10.12, and it can be evaluated by the CVGM, implying that IH correlated damage follows the damage rule of the CVGM.

According to the above assumptions, the following damage evolution rule is proposed:

$$\mathrm{d}D = \mathrm{d}D_{\mathrm{IH}} + \mathrm{d}D_{\mathrm{KH}} \tag{10.1}$$

$$\mathrm{d}D_{\mathrm{IH}} = \mathrm{d}D_{\mathrm{CVGM}} \tag{10.2}$$

$$\mathrm{d}D_{\mathrm{KH}} = \eta \mathrm{d}D_{\mathrm{IH}} \tag{10.3}$$

where D is the total damage, D_{IH} and D_{KH} are, respectively, the IH and KH correlated damages. D_{CVGM} is the damage index defined by the CVGM, and η is a material constant. The same as that of the CVGM, crack initiation is postulated to occur when D reaches 1.0.

According to Eqs. (10.1)–(10.3), one can evaluate crack initiation of ductile metallic specimens under ULCF loading with the assistance of the previously proposed CVGM, where there is an additional work is to distinguish D_{KH} from D_{IH}. Decomposition of D_{IH} and D_{KH} for the node with the maximum damage index according to numerical results of the DEN specimen using the CVGM is illustrated in Fig. 10.13. The diagonal segments shown in the curve of Fig. 10.13 represent the tensile loading half cycles of the node, and the horizontal segments are the compressive loading half cycles where no damage accumulates according to the CVGM. It can be found from Fig. 10.13 that the damage during the first tensile loading half cycle is 100% D_{IH} and that of the second tensile loading half cycle consists of both D_{IH} and D_{KH}, since IH occurs during the second tensile loading half cycle according to the definition of IH as illustrated in Fig. 10.12. D_{IH} can be identified by comparison of the equivalent plastic strain increment between the first and second tensile loading half cycles. For the subsequent loading half cycles except for the first two loading cycles, only D_{KH} accumulates since only KH occurs according to the definition illustrated in Fig. 10.12. Based on the above discussions, all the damage increment of the D_{CVGM} can be classified into D_{IH} and D_{KH}, and the total damage according to Eq. (10.1) can be obtained.

Based on this model, the crack initiation life of ductile metal under ULCF loading with constant-amplitude can be correlated with monotonic tension coupon test results according to the flowchart shown in Fig. 10.14, which can be divided into the following steps:

Step 1: Conduct coupon tests under monotonic tension.

Step 2: Calibrate the material parameter χ_{cr} using numerical analysis of the coupon test results.

Step 3: Carry out numerical analysis of the specimens under the first three loading cycles to determine the values of D_{IH} and D_{KH} using the CVGM, and the

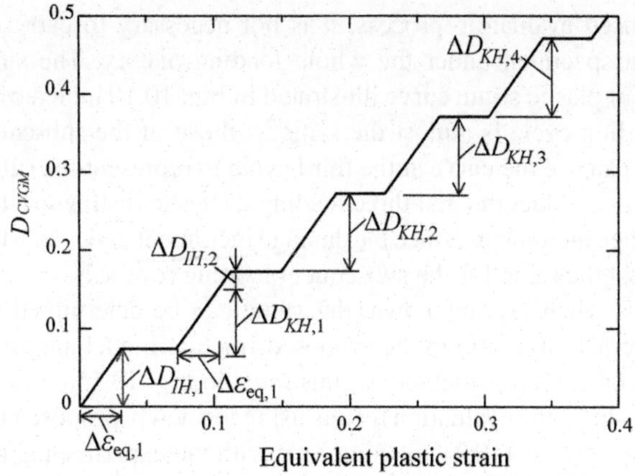

Fig. 10.13 Decomposition of isotropic hardening and kinematic hardening correlated damage

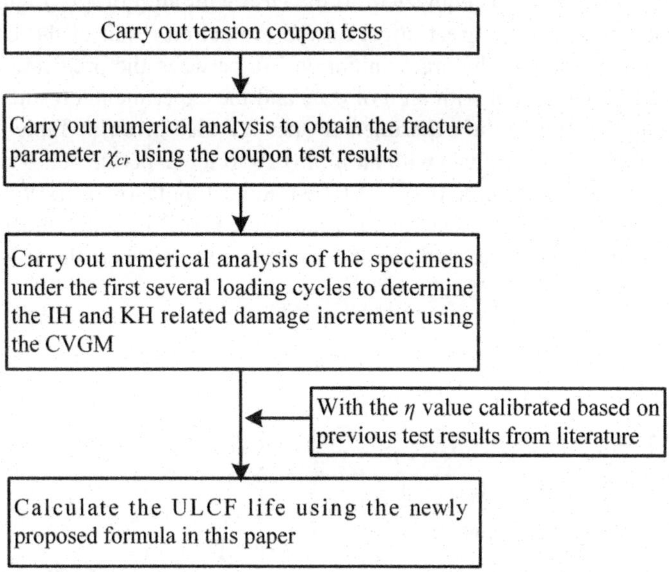

Fig. 10.14 Flowchart to evaluate ultra-low-cycle fatigue life using tension coupon test and finite element analysis

damage under the subsequent loading cycles is all KH correlated with the same value as that of the third loading cycle.

Step 4: Postulating that the material constant, η, in Eq. (10.3) is known from previous studies, the crack initiation life of a specimen can be predicted at the instant when the sum of D_{IH} and D_{KH} reaches 1.0.

In the proposed evaluation process, it is not necessary to carry out numerical simulation of a specimen under the whole loading history. The stress triaxiality versus equivalent plastic strain curve illustrated in Fig. 10.10 indicates that the curve at the third loading cycle is almost the same as those of the subsequent reversals. Therefore, one can use the curve at the third cycle to represent the subsequent ones. One just needs to conduct the first three loading cycles to distinguish the IH and KH correlated damage increments, since the damage increment under constant-amplitude loading is almost the same for the subsequent loading reversals.

For the cases when η is unknown, the value can be determined by comparing the instants of crack initiation in the proposed fracture model and those of the test results by the trial and error method as illustrated in Fig. 10.15a. For the aluminum 6061-T6 in this study, the evaluation results using the newly proposed fracture model according to Eqs. (10.1)–(10.3) are compared with the experimental results and the analysis results of the CVGM in Fig. 10.15 with η ranging from 0.1 to 0.9, where the value of 0.29 fits best for the experimental results of the DEN specimens. It should be noted that the CVGM is equivalent to the case with an η of 1.0. Figure 10.15a also indicates that η has a great effect on the evaluation result of the ULCF life. The comparison result of the crack initiation life between the predicted results of the proposed ULCF model with an η of 0.29 and the experimental results, and that between the CVGM and the experimental ones are also given in Table 10.4. The average of the evaluated values with an η of 0.29 to the experimental ones is 1.01, and the coefficient of variation (CoV) is 0.094, indicating favorable performance of the proposed fracture model. On the other hand, the CVGM greatly underestimates crack initiation life compared with the experimental results, where the average value is 0.36, and the CoV is 0.086.

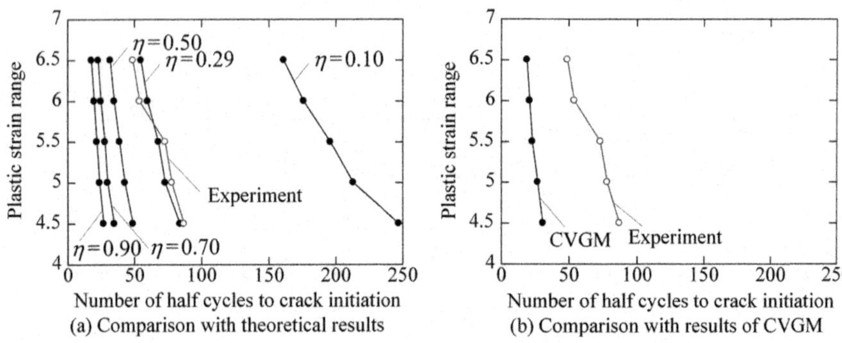

Fig. 10.15 Comparison of number of half cycles till crack initiation

10.5 Summaries

In this study, a new fracture model for evaluating ultra-low-cycle fatigue life (ULCF) of ductile metals is proposed, in which the different accumulating rates of the isotropic hardening (IH) and kinematic hardening (KH) correlated damage is considered. The model can be employed to correlate and evaluate number of loading cycles till crack initiation of ductile metals only using monotonic tension coupon test results. The proposed fracture model and evaluation approach are validated through ULCF tests on 15 double-edge-notched (DEN) specimens made of aluminum alloy 6061-T6. The evaluation results of the newly proposed fracture model are also compared with those of a previously proposed cyclic void growth model (CVGM). Fractographic study using a scanning electron microscope (SEM) was also conducted to verify the fracture modes of the DEN specimens. Based on the experimental and numerical studies, the following conclusions can be drawn:

(1) Ductility and ductile cracking resistance of the aluminum alloy 6061-T6 are lower than those of common mild steel.
(2) The ULCF surfaces of the DEN specimens consists of two distinguished parts, i.e., ULCF surface and ductile fracture surface, which are, respectively, induced by the cyclic loading reversals and final rupture.
(3) The size of the dimples under SEM observation ranges from 10 to 100 μm, which is much less than that of common mild steel.
(4) The ULCF surface has only several visible striations with much wider width compared with those of high- and low-cycle fatigue problems. In addition, the fractographic observation results show that microstructure on the ULCF striation is typical dimple pattern and that within two ULCF striations is quasi-cleavage fracture surface mixed with flattened ductile dimples.
(5) The CVGM greatly underestimates the number of loading cycles till crack initiation for the cases under constant-amplitude loading with a ULCF life of dozens of loading cycles, while the newly proposed model can evaluate the crack initiation instants well.
(6) The proposed model postulates a linear relationship between the IH and KH correlated damage, and the ratio of the KH correlated damage accumulation rate to that of the IH is 0.29 for the aluminum alloy 6061-T6 in this study.

It should be noted that the ratio, 0.29, of the KH correlated damage accumulation rate to that of the IH is obtained from the experimental and numerical results for the aluminum alloy 6061-T6. The value may be dependent on material type, loading history, and stress states. Therefore, more experimental and numerical analyses are required to obtain the correlation between the IH and KH correlated damage for other ductile metal under a variety of loading conditions in future studies. The newly proposed fracture model coupling with proper cyclic plasticity rules should be developed to automatically distinguish the IH and KH correlated damage, which will further simplify the evaluation process in this study.

References

Al Zamzami I, Susmel L (2017) On the accuracy of nominal, structural, and local stress based approaches in designing aluminium welded joints against fatigue. Int J Fatigue 101(Part 2):137–158

Ambriz RR et al (2010) Effect of the welding profile generated by the modified indirect electric arc technique on the fatigue behavior of 6061-T6 aluminum alloy. Mater Sci Eng, A 527:2057–2064

Bai Y, Wierzbicki T (2015) A comparative study of three groups of ductile fracture loci in the 3D space. Eng Fract Mech 135:147–167

Bao Y, Wierzbicki T (2005) On the cut-off value of negative triaxiality for fracture. Eng Fract Mech 72:1049–1069

González R et al (2011) Experimental investigation for fatigue strength of a cast aluminium alloy. Int J Fatigue 33:273–278

Hao H et al (2014) Strain ratio effects on low-cycle fatigue behavior and deformation microstructure of 2124-T851 aluminum alloy. Mater Sci Eng, A 605:151–159

Jia L-J, Kuwamura H (2014) Ductile fracture simulation of structural steels under monotonic tension. J Struct Eng (ASCE) 140:04013115

Jia L-J, Kuwamura H (2015) Ductile fracture model for structural steel under cyclic large strain loading. J Constr Steel Res 106:110–121

Jia L-J et al (2014) Experimental and numerical study of postbuckling ductile fracture of heat-treated SHS stub columns. J Struct Eng (ASCE) 140:04014044

Jia L-J et al (2016a) Experimental and numerical study on ductile fracture of structural steels under combined shear and tension. J Bridge Eng (ASCE) 21(5):04016008

Jia L-J et al (2016b) Ductile crack initiation and propagation of structural steels under cyclic combined shear and normal stress loading. Constr Build Mater 112:69–83

Khandelwal K, El-Tawil S (2014) A finite strain continuum damage model for simulating ductile fracture in steels. Eng Fract Mech 116:172–189

Kiran R, Khandelwal K (2013) A micromechanical model for ductile fracture prediction in ASTM A992 steels. Eng Fract Mech 102:101–117

Kiran R, Khandelwal K (2014) A triaxiality and Lode parameter dependent ductile fracture criterion. Eng Fract Mech 128:121–138

Kuroda M (2002) Extremely low cycle fatigue life prediction based on a new cumulative fatigue damage model. Int J Fatigue 24:699–703

Kuwamura H (1997) Transition between fatigue and ductile fracture in steel. J Struct Eng (ASCE) 123:864–870

Liu Y et al (2017) Ductile-fatigue transition fracture mode of welded T-joints under quasi-static cyclic large plastic strain loading. Eng Fract Mech 176:38–60

May A et al (2013) Analysis of the cyclic behavior and fatigue damage of extruded AA2017 aluminum alloy. Mater Sci Eng, A 571:123–136

Naeimi M et al (2017) Correlation between microstructure, tensile properties and fatigue life of AA1050 aluminum alloy processed by pure shear extrusion. Mater Sci Eng, A 679:292–298

Ohata M, Toyoda M (2004) Damage concept for evaluating ductile cracking of steel structure subjected to large-scale cyclic straining. Sci Technol Adv Mater 5:241–249

Özdeş H, Tiryakioğlu M (2017) On estimating high-cycle fatigue life of cast Al-Si-Mg-(Cu) alloys from tensile test results. Mater Sci Eng, A 688:9–15

Pang JC et al (2013) General relation between tensile strength and fatigue strength of metallic materials. Mater Sci Eng, A 564:331–341

Panontin TL, Sheppard SD (1995) The relationship between constraint and ductile fracture initiation as defined by micromechanical analyses. ASTM STP 1256, ASTM, West Conshohoken, PA, pp 54–85

Xue L, Wierzbicki T (2008) Ductile fracture initiation and propagation modeling using damage plasticity theory. Eng Fract Mech 75:3276–3293

Yoshida F, Uemori T (2002) A model of large-strain cyclic plasticity describing the Bauschinger effect and workhardening stagnation. Int J Plast 18:661–686

Chapter 11
Summary and Future Study

11.1 Main Conclusion

The applicability of several classical plasticity models to ultra-low-cycle fatigue loading of structural steels and aluminum is validated, and a new plasticity model for large strain reversals is proposed. Ductile fracture models for cyclic incremental and constant-amplitude loading are also proposed. Approaches to calibrate the model parameters of the plasticity models and the fracture models only from monotonic tension coupon tests are proposed. The models with the calibration methods are validated by a series of cyclic tests on hourglass-type specimens and applied to structural steels, steel connections, structural aluminum, and aluminum buckling-restrained braces successfully. The detailed findings are given in the following.

11.1.1 Cyclic Plasticity Models for Cyclic Large Strain Loading

Based on the experimental and numerical studies on a series of hourglass-type steel specimens and double-edge-notched aluminum specimens, the following main conclusions can be drawn:

(1) A modified weighted average method is proposed to obtain the true stress–true strain till fracture from the test result of a tension coupon.
(2) A modified Yoshida–Uemori model is proposed, and the model is found to be able to trace the cyclic plasticity of structural steels at very large plastic strain ranges. The model parameters can also be obtained from a tension coupon test by carrying out an optimization analysis.
(3) The Chaboche model with isotropic hardening (IH) can well simulate the cyclic plasticity of structural steels under cyclic large strain loading, and the model

L.-J. Jia and H. Ge, *Ultra-low-Cycle Fatigue Failure of Metal Structures under Strong Earthquakes*, Springer Tracts in Civil Engineering,
https://doi.org/10.1007/978-981-13-2661-5_11

slightly overestimates the stress for the case under constant-amplitude loading due to a lack of the memory surface.

(4) A simple method using representative mechanical parameters of aluminum is proposed to obtain the true stress–true strain till fracture, and the proposed method can be applied to evaluating hysteretic properties of aluminum coupons and buckling-restrained braces under cyclic loading with good accuracy.

11.1.2 Ductile Fracture Models for Cyclic Large Strain Loading

(1) A monotonic fracture model with only one model parameter for structural steels is proposed by applying the Rice–Tracey model and the Miner's rule in incremental form.
(2) A crack propagation rule based on the energy concept is proposed for structural steels, and the method performs well for high-strength steel, where the correlated parameter can be obtained through monotonic tension coupon test.
(3) A cyclic fracture model developed from the monotonic fracture model in combination with the Chaboche model with IH, or the modified Yoshida–Uemori model, can generally well simulate the ductile fracture of structural steels under cyclic large strain loading.
(4) Experimental tests on stub box columns under ULCF loading are carried out to calibrate both the cyclic fracture model and the Chaboche model with IH. The numerical simulation results using the models agree well with the corresponding test results.
(5) Ultra-low-cycle fatigue (ULCF) loading tests on structural steel and aluminum are conducted, where a transition fracture mode from low-cycle fatigue to ductile fracture is observed.
(6) An approach to evaluate the ULCF life of ductile metals is proposed, which can be correlated with the corresponding tension coupon test results. The theory has been validated using aluminum coupons under ULCF loading.

11.2 Future Work

It is suggested that the following items require further study on ductile fracture of metals under ULCF loading.

(1) The damage rule within negative stress triaxialities under cyclic loading at large plastic ranges close to fracture of metals is found to be underestimated in this study, and a more accurate rule is required based on more experimental results.

(2) Ductile crack propagation simulation of cracked bodies under cyclic loading should be further studied, since the final sudden loss for specimens under ULCF loading sometimes cannot be well captured.
(3) Effects of residual stress and strain on buckling modes and fracture processes of steel members need further study.
(4) Though the transition between brittle fracture and ductile fracture is not the concern of this study, it is of great importance in practice, since the brittle fracture is found to be triggered by ductile fracture. It is still an open issue and needs further study.
(5) The tests carried out in this study is only on structural steels and aluminum, and the validity of the fracture model and the plasticity models for other ductile metals should be further validated.
(6) The welded structures are still difficult to be predicted due to the complicated material properties within the heat-affected zone, and a number of factors affect the mechanical properties and defects of the welds. There are still no straightforward and accurate approaches to evaluate the welded structures, where further study is necessary.
(7) The effect of strain rate on ductile fracture of structural steels has not been comprehensively studied yet. Ductile fracture mechanisms of structural steels under dynamic ULCF loading require further study.
(8) Study on seismic performance of large-scale whole structures has seldom employed such elaborate plasticity models and fracture models as used in this study. Ductile fracture simulation of whole structures under seismic loading requires further study, and performance of metallic structures at the ultimate failure stage can be re-evaluated to ensure a safer design.
(9) More elaborate simulation of the cracking process is necessary, e.g., closure and contact of the neighboring cracked parts; more realistic simulation of the cracking path using advanced approaches such as the XFEM, while convergence and simulation efficiency are also greatly concerned for large-scaled structures.

Appendix A
Implementation of the Modified Yoshida–Uemori Model

Abstract Rate-independent metal plasticity models are widely applied to many fields, where most of the models are only effective for small strain ranges. A modified Yoshida–Uemori model was previously proposed and validated, which can well evaluate metal plasticity at the full strain range till fracture. The model is difficult to implement into finite element software due to its complicated coupled formation for isotropic and kinematic hardening, which greatly limits its application. In this chapter, the modified Yoshida–Uemori model is implemented using a robust integration algorithm termed as the adaptive substepping method. The implemented model is further validated by cyclic tests on mild steels at both material and member levels.

A.1 Introduction

A new two-surface model with non-isotropic hardening memory surface (Ohno 1982) was proposed and termed the Yoshida–Uemori model (Yoshida and Uemori 2002), which was reported to perform well for the simulation of cyclic metal plasticity at large plastic strain ranges until around 20%. However, the model has some limitations at extremely large plastic strain ranges (Jia and Kuwamura 2014; Shi et al. 2008), especially for the studies on ductile fracture and sheet metal forming, where extremely large plastic strain exceeding 100% will be of interest. A modified Yoshida–Uemori model was thus proposed considering the limitations of the model at extremely large plastic strain ranges, which can well simulate metal plasticity of mild steel up to fracture.

For the simulation of ductile fracture, the equivalent plastic strain at fracture of metals can reach a value over 100% (Bao and Wierzbicki 2004; Bao and Wierzbicki 2005; Jia and Kuwamura 2014; Wierzbicki et al. 2005). Under this situation, it is

L. Jia and H. Ge, *Ultra-low-Cycle Fatigue Failure of Metal Structures under Strong Earthquakes*, Springer Tracts in Civil Engineering,
https://doi.org/10.1007/978-981-13-2661-5

difficult to incorporate a cyclic plasticity model with complicated formation. In extremely large plastic strain ranges, integration algorithms become critical, since numerical analyses will be difficult to converge due to the high geometrical and material nonlinearity, especially for the Yoshida–Uemori model with three coupled backstresses, a bounding surface with isotropic hardening, and a memory surface. For the implementation of a complicated model, commonly finite element (FE) software requires a material subroutine to provide a stiffness matrix to map a small strain perturbation to a corresponding perturbed stress, where the stiffness matrix is commonly called the consistent tangent matrix (Crisfield et al. 2012; Dunne and Petrinic 2005), which is of great importance to ensure rapid convergence of a subroutine. The original Yoshida–Uemori model was implemented using the backward Euler method for the application to prediction of springback problems (Ghaei and Green 2010; Ghaei et al. 2010). This study aims to implement the modified Yoshida–Uemori model to apply the model to extremely large plastic strain problems, where the convergence of the implementation is a main concern due to the highly nonlinear coupling between the three internal backstresses and the corresponding isotropic hardening components, especially at extremely large plastic strain ranges, where severe strain concentration occurs. Possible approaches were proposed to solve this kind of problems, e.g., utilizing specific auxiliary predictor surfaces (Bićanić and Pearce 1996) or the line search schemes (Pérez-Foguet and Armero 2002). Another approach to ensure the convergence of the global iterations is the substepping method (Abbo and Sloan 1996; Pérez-Foguet et al. 2001; Sloan 1987), where an incremental step is further divided into several substeps, and the integration rule in each substep is the same as that of a single-step integration rule. Herein, the consistent tangent matrices of the modified Yoshida–Uemori using both the single-step and substepping integration methods are deducted, and the integration formulae of the memory surface in the strain space are also derived. Meanwhile, a method to estimate the optimal number of substeps considering both convergence and computational efficiency is proposed. Finally, the subroutine was successfully implemented into ABAQUS, and the calibration analyses of the subroutine are carried out by comparison between numerical and theoretical results. The model is also applied to simulate cyclic behaviors of structural steels and steel members in extremely large plastic strain ranges, where the numerical results are in favorable agreement with the experimental ones.

A.2 Stress Integration of the Single-Step Integration Method

A.2.1 Algorithm of Stress Integration

For computational plasticity, the key problem that is related with the constitutive models is to obtain the current values of state variables such as stress, strain and internal variables by integrating the local constitutive equations under certain initial conditions. For most of the occasions, the main problem is to calculate the current values of state variables (e.g., stress $\boldsymbol{\sigma}_{n+1}$, plastic strain $\boldsymbol{\varepsilon}_{p,\,n+1}$, and backstress $\boldsymbol{\alpha}_{n+1}$, $\boldsymbol{\beta}_{n+1}$, $\boldsymbol{\theta}_{n+1}$) based on the given values of the state variables at the nth increment (e.g., stress $\boldsymbol{\sigma}_n$, plastic strain $\boldsymbol{\varepsilon}_{p,\,n}$, and the backstresses $\boldsymbol{\alpha}_n$, $\boldsymbol{\beta}_n$, $\boldsymbol{\theta}_n$) and the strain increment at the $(n + 1)$th increment, $\Delta\boldsymbol{\varepsilon}_{n+1}$.

In the present implementation of the modified Yoshida–Uemori model, an implicit backward Euler difference scheme is employed to integrate the constitutive equations, which is geometrically interpreted as the closest point projection in the energy norm of trial elastic state onto the elastic domain (Simo and Hughes 1998) as shown in Fig. A.1.

The closest point projection method includes an elastic prediction step and a plastic modification step as shown in the figure. The numerical scheme mainly consists of stress integration, which transforms a series of constitutive equations into a system of nonlinear equations and the solution into the system of nonlinear equations, where the backward Euler difference method is adopted.

For the closest point projection method, the trial stress at the $(n + 1)$th increment, $\boldsymbol{\sigma}^{\mathrm{trial}}$, is first obtained based on elastic evaluation

$$\boldsymbol{\sigma}_{n+1}^{\mathbf{trial}} = \boldsymbol{\sigma}_n + \boldsymbol{D}:\mathrm{d}\boldsymbol{\varepsilon}_{n+1} \tag{A.1}$$

where $\boldsymbol{\sigma}_n$ is the stress at the nth increment, $\boldsymbol{D}$ is the elasticity tensor, and $\mathrm{d}\boldsymbol{\varepsilon}_{n+1}$ is strain increment at the $(n + 1)$th increment. For the following sections, the subscript $n + 1$ will be neglected for simplicity, and the variables without subscripts denote the corresponding variables at the $(n + 1)$th increment.

Then in the plastic modification step, the current stress is obtained by

$$\boldsymbol{\sigma} = \boldsymbol{\sigma}_n + \boldsymbol{D}:(\mathrm{d}\boldsymbol{\varepsilon} - \mathrm{d}\boldsymbol{\varepsilon}_p) = \boldsymbol{\sigma}_n + \boldsymbol{D}:\mathrm{d}\boldsymbol{\varepsilon} - \boldsymbol{D}:\mathrm{d}\boldsymbol{\varepsilon}_p = \boldsymbol{\sigma}^{\mathbf{trial}} - \boldsymbol{D}:\mathrm{d}\boldsymbol{\varepsilon}_p \tag{A.2}$$

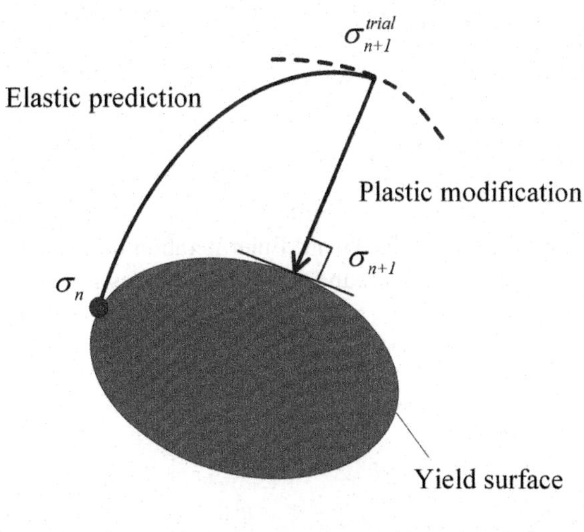

Fig. A.1 Illustration of closest point projection for single-step integration method

Equation (A.2) can be given as

$$\mathrm{d}\boldsymbol{\varepsilon}_p = \boldsymbol{D}^{-1}:\mathrm{d}\boldsymbol{\sigma} \tag{A.3}$$

If the plastic strain increment is obtained, then the stress and other state variables can also be calculated based on the constitutive equations.

For the current implementation, the associated flow rule is assumed. Then, the plastic strain increment can be given as

$$\mathrm{d}\boldsymbol{\varepsilon}_p = \mathrm{d}\varepsilon_{\mathrm{eq}}\frac{\partial f}{\partial \boldsymbol{\sigma}} = \mathrm{d}\varepsilon_{\mathrm{eq}}\cdot \boldsymbol{n} \tag{A.4}$$

where $\boldsymbol{n}$ is the normal to the yield surface.

According to Eqs. (3.15), (3.18), and (A.4), the stress integration equations of the modified Yoshida–Uemori model can be given as

$$\boldsymbol{F}_1 = -\boldsymbol{\varepsilon}_p + \boldsymbol{\varepsilon}_{p,n} + \mathrm{d}\varepsilon_{\mathrm{eq}}\cdot \boldsymbol{n} = 0 \tag{A.5}$$

$$\boldsymbol{F}_2 = -\boldsymbol{\theta} + \boldsymbol{\theta}_{\mathrm{n}} + \boldsymbol{H}_1\mathrm{d}\varepsilon_{\mathrm{eq}} = 0 \tag{A.6}$$

$$\boldsymbol{F}_3 = -\boldsymbol{\beta} + \boldsymbol{\beta}_{\mathrm{n}} + \boldsymbol{H}_2\mathrm{d}\varepsilon_{\mathrm{eq}} = 0 \tag{A.7}$$

where

$$\boldsymbol{H}_1 = C\left[\frac{a}{\sigma^{\mathrm{trial}}}(\boldsymbol{S}-\boldsymbol{\alpha}) - \sqrt{\frac{a}{\bar{\theta}}}\boldsymbol{\theta}\right] \tag{A.8}$$

$$\boldsymbol{H}_2 = m\left(b\frac{\boldsymbol{S}-\boldsymbol{\alpha}}{\sigma^{\mathrm{trial}}} - \boldsymbol{\beta}\right) \tag{A.9}$$

where $\sigma^{\mathrm{trial}} = \sqrt{3/2\boldsymbol{\sigma}^{\mathbf{trial}}:\boldsymbol{\sigma}^{\mathbf{trial}}}$.

The consistency condition of the yield function, Eq. (3.11), is also necessary to be included as

$$F_4 = \sqrt{\frac{3}{2}(\boldsymbol{S}-\boldsymbol{\alpha}):(\boldsymbol{S}-\boldsymbol{\alpha})} - \sigma_{y0} = 0 \tag{A.10}$$

To solve the system of equations above, the equations have to be linearized and solved using the backward Euler iteration method. The linearization of Eqs. (A.5)–(A.7) and (A.10) around the state variables gives

$$\boldsymbol{F}_1^{(\mathrm{k})} + \boldsymbol{D}^{-1}:\mathrm{d}\boldsymbol{\sigma}^{(\mathrm{k})} + \mathrm{d}\varepsilon_{\mathrm{eq}}^{(\mathrm{k})}\mathrm{d}\boldsymbol{n}^{(\mathrm{k})} + \delta(\mathrm{d}\varepsilon_{\mathrm{eq}}^{(\mathrm{k})})\cdot \boldsymbol{n}^{(\mathrm{k})} = 0 \tag{A.11}$$

$$\boldsymbol{F}_2 = -\boldsymbol{\theta} + \boldsymbol{\theta}_{\mathrm{n}} + \boldsymbol{H}_1 \mathrm{d}\varepsilon_{\mathrm{eq}} = 0 \tag{A.12}$$

$$\boldsymbol{F}_3^{(\mathrm{k})} - \mathrm{d}\boldsymbol{\beta}^{(\mathrm{k})} + \mathrm{d}\varepsilon_{\mathrm{eq}}^{(\mathrm{k})} \mathrm{d}\boldsymbol{H}_2^{(\mathrm{k})} + \delta(\mathrm{d}\varepsilon_{\mathrm{eq}}^{(\mathrm{k})}) \boldsymbol{H}_2^{(\mathrm{k})} = 0 \tag{A.13}$$

$$F_4^{(\mathrm{k})} + \boldsymbol{f}_{\sigma}^{(\mathrm{k})} : \mathrm{d}\boldsymbol{\sigma}^{(\mathrm{k})} + \boldsymbol{f}_{\theta}^{(\mathrm{k})} : \mathrm{d}\boldsymbol{\theta}^{(\mathrm{k})} + \boldsymbol{f}_{\beta}^{(\mathrm{k})} : \mathrm{d}\boldsymbol{\beta}^{(\mathrm{k})} = 0 \tag{A.14}$$

where

$$\mathrm{d}\boldsymbol{n}^{(\mathrm{k})} = \boldsymbol{n}_{\sigma}^{(\mathrm{k})} : \mathrm{d}\boldsymbol{\sigma}^{(\mathrm{k})} + \boldsymbol{n}_{\theta}^{(\mathrm{k})} : \mathrm{d}\boldsymbol{\theta}^{(\mathrm{k})} + \boldsymbol{n}_{\beta}^{(\mathrm{k})} : \mathrm{d}\boldsymbol{\beta}^{(\mathrm{k})} \tag{A.15}$$

$$\mathrm{d}\boldsymbol{H}_1^{(\mathrm{k})} = \boldsymbol{H}_{1\sigma}^{(\mathrm{k})} : \mathrm{d}\boldsymbol{\sigma}^{(\mathrm{k})} + \boldsymbol{H}_{1\theta}^{(\mathrm{k})} : \mathrm{d}\boldsymbol{\theta}^{(\mathrm{k})} + \boldsymbol{H}_{1\beta}^{(\mathrm{k})} : \mathrm{d}\boldsymbol{\beta}^{(\mathrm{k})} + \mathrm{d}\varepsilon_{\mathrm{eq}}^{(\mathrm{k})} \boldsymbol{H}_{1\mathrm{d}\varepsilon_{\mathrm{eq}}}^{(\mathrm{k})} \tag{A.16}$$

$$\mathrm{d}\boldsymbol{H}_2^{(\mathrm{k})} = \boldsymbol{H}_{2\sigma}^{(\mathrm{k})} : \mathrm{d}\boldsymbol{\sigma}^{(\mathrm{k})} + \boldsymbol{H}_{2\theta}^{(\mathrm{k})} : \mathrm{d}\boldsymbol{\theta}^{(\mathrm{k})} + \boldsymbol{H}_{2\beta}^{(\mathrm{k})} : \mathrm{d}\boldsymbol{\beta}^{(\mathrm{k})} \tag{A.17}$$

Then, one can obtain

$$\left[\boldsymbol{A}^{(\mathrm{k})}\right]^{-1} \begin{Bmatrix} \mathrm{d}\boldsymbol{\sigma}^{(\mathrm{k})} \\ \mathrm{d}\boldsymbol{\theta}^{(\mathrm{k})} \\ \mathrm{d}\boldsymbol{\beta}^{(\mathrm{k})} \end{Bmatrix} = -\left\{\widetilde{\boldsymbol{F}}^{(\mathrm{k})}\right\} - \delta(\mathrm{d}\varepsilon_{\mathrm{eq}}^{(\mathrm{k})}) \left\{\tilde{\boldsymbol{n}}^{(\mathrm{k})}\right\} - \mathrm{d}\varepsilon_{\mathrm{eq}}^{2(\mathrm{k})} \left\{\widetilde{\boldsymbol{H}}_{1d\varepsilon_{\mathrm{eq}}}^{(\mathrm{k})}\right\} \tag{A.18}$$

where

$$\left[\boldsymbol{A}^{(\mathrm{k})}\right]^{-1} = \begin{bmatrix} \boldsymbol{D}^{-1} + \mathrm{d}\varepsilon_{\mathrm{eq}}^{(\mathrm{k})} \boldsymbol{n}_{\sigma}^{(\mathrm{k})} & \mathrm{d}\varepsilon_{\mathrm{eq}}^{(\mathrm{k})} \boldsymbol{n}_{\theta}^{(\mathrm{k})} & \mathrm{d}\varepsilon_{\mathrm{eq}}^{(\mathrm{k})} \boldsymbol{n}_{\beta}^{(\mathrm{k})} \\ \mathrm{d}\varepsilon_{\mathrm{eq}}^{(\mathrm{k})} \boldsymbol{H}_{1\sigma}^{(\mathrm{k})} & -\boldsymbol{I} + \mathrm{d}\varepsilon_{\mathrm{eq}}^{(\mathrm{k})} \boldsymbol{H}_{1\theta}^{(\mathrm{k})} & \mathrm{d}\varepsilon_{\mathrm{eq}}^{(\mathrm{k})} \boldsymbol{H}_{1\beta}^{(\mathrm{k})} \\ \mathrm{d}\varepsilon_{\mathrm{eq}}^{(\mathrm{k})} \boldsymbol{H}_{2\sigma}^{(\mathrm{k})} & \mathrm{d}\varepsilon_{\mathrm{eq}}^{(\mathrm{k})} \boldsymbol{H}_{2\theta}^{(\mathrm{k})} & -\boldsymbol{I} + \mathrm{d}\varepsilon_{\mathrm{eq}}^{(\mathrm{k})} \boldsymbol{H}_{2\beta}^{(\mathrm{k})} \end{bmatrix} \tag{A.19}$$

where $\boldsymbol{n}_{\sigma}$, $\boldsymbol{n}_{\theta}$, $\boldsymbol{n}_{\beta}$, $\boldsymbol{H}_{1\sigma}$, $\boldsymbol{H}_{1\theta}$, $\boldsymbol{H}_{1\beta}$, $\boldsymbol{H}_{2\sigma}$, $\boldsymbol{H}_{2\theta}$, $\boldsymbol{H}_{2\beta}$ are the differences of $\boldsymbol{n}$, $\boldsymbol{H}_1$, and $\boldsymbol{H}_2$ with respect to $\boldsymbol{\sigma}$, $\boldsymbol{\theta}$, and $\boldsymbol{\beta}$.

$$\left\{\widetilde{\boldsymbol{F}}^{(\mathrm{k})}\right\} = \begin{Bmatrix} \boldsymbol{F}_1^{(\mathrm{k})} \\ \boldsymbol{F}_2^{(\mathrm{k})} \\ \boldsymbol{F}_3^{(\mathrm{k})} \end{Bmatrix}, \left\{\tilde{\boldsymbol{n}}^{(\mathrm{k})}\right\} = \begin{Bmatrix} \boldsymbol{n}^{(\mathrm{k})} \\ \boldsymbol{H}_1^{(\mathrm{k})} \\ \boldsymbol{H}_2^{(\mathrm{k})} \end{Bmatrix}$$

$$\text{when } (\boldsymbol{\beta} - \boldsymbol{q}) : \mathrm{d}\boldsymbol{\beta} > 0 \left\{\widetilde{\boldsymbol{H}}_{1\mathrm{d}\varepsilon_{\mathrm{eq}}}^{(\mathrm{k})}\right\} = \begin{Bmatrix} 0 \\ \boldsymbol{H}_{1\mathrm{d}\varepsilon_{\mathrm{eq}}}^{(\mathrm{k})} \\ 0 \end{Bmatrix} \quad \text{otherwise} \left\{\widetilde{\boldsymbol{H}}_{1\mathrm{d}\varepsilon_{\mathrm{eq}}}^{(\mathrm{k})}\right\} = \{0\} \tag{A.20}$$

Then, the increments of stress and internal state variables can be given as

$$\begin{Bmatrix} \mathrm{d}\boldsymbol{\sigma}^{(\mathrm{k})} \\ \mathrm{d}\boldsymbol{\theta}^{(\mathrm{k})} \\ \mathrm{d}\boldsymbol{\beta}^{(\mathrm{k})} \end{Bmatrix} = -\left[\boldsymbol{A}^{(\mathrm{k})}\right]:\left\{\widetilde{\boldsymbol{F}}^{(\mathrm{k})}\right\} - \delta \mathrm{d}\varepsilon_{\mathrm{eq}}^{(\mathrm{k})}\left[\boldsymbol{A}^{(\mathrm{k})}\right]:\left\{\tilde{\boldsymbol{n}}^{(\mathrm{k})}\right\} - \left(\mathrm{d}\varepsilon_{eq}^{(\mathrm{k})}\right)^2\left[\boldsymbol{A}^{(\mathrm{k})}\right] :\left\{\widetilde{\mathbf{H}}_{\mathbf{1}\mathrm{d}\varepsilon_{\mathrm{eq}}}^{(\mathrm{k})}\right\} \tag{A.21}$$

Substitute Eq. (A.21) into Eq. (A.14), one can obtain

$$\delta\left(\mathrm{d}\varepsilon_{\mathrm{eq}}^{(\mathrm{k})}\right) = \frac{F_4^{(\mathrm{k})} - \partial f^{(\mathrm{k})}:\left[\boldsymbol{A}^{(\mathrm{k})}\right]:\left\{\widetilde{\boldsymbol{F}}^{(\mathrm{k})}\right\} - \left(\mathrm{d}\varepsilon_{\mathrm{eq}}^{(\mathrm{k})}\right)^2 \partial f:\left[\boldsymbol{A}^{(\mathrm{k})}\right]:\left\{\widetilde{\boldsymbol{H}}_{\mathbf{1}\mathrm{d}\varepsilon_{\mathrm{eq}}}^{(\mathrm{k})}\right\}}{\partial f^{(\mathrm{k})}:\left[\boldsymbol{A}^{(\mathrm{k})}\right]:\left\{\tilde{\boldsymbol{n}}^{(\mathrm{k})}\right\}} \tag{A.22}$$

$$\delta\left(\mathrm{d}\varepsilon_{\mathrm{eq}}^{(\mathrm{k})}\right) = \frac{\partial f{:}\boldsymbol{A}^{(\mathrm{k})}:\begin{Bmatrix} \mathrm{d}\boldsymbol{\varepsilon}^{(\mathrm{k})} \\ 0 \\ 0 \end{Bmatrix} - \left(\mathrm{d}\varepsilon_{\mathrm{eq}}^{(\mathrm{k})}\right)^2 \partial f{:}\boldsymbol{A}^{(\mathrm{k})}:\begin{Bmatrix} 0 \\ \boldsymbol{H}_{\mathbf{1}\mathrm{d}\varepsilon_{\mathrm{eq}}}^{(\mathrm{k})} \\ 0 \end{Bmatrix}}{\partial f{:}\boldsymbol{A}^{(\mathrm{k})}:\left\{\tilde{\boldsymbol{n}}^{(\mathrm{k})}\right\}} \tag{A.23}$$

Substitute Eq. (A.23) into Eq. (A.18), one can finally obtain increments of stress and internal state variables

$$\begin{Bmatrix} \mathrm{d}\boldsymbol{\sigma}^{(\mathrm{k})} \\ \mathrm{d}\boldsymbol{\theta}^{(\mathrm{k})} \\ \mathrm{d}\boldsymbol{\beta}^{(\mathrm{k})} \end{Bmatrix} = \boldsymbol{A}^{(\mathrm{k})}:\begin{Bmatrix} \mathrm{d}\boldsymbol{\varepsilon}^{(\mathrm{k})} \\ 0 \\ 0 \end{Bmatrix} - \left(\mathrm{d}\varepsilon_{\mathrm{eq}}^{(\mathrm{k})}\right)^2 \boldsymbol{A}^{(\mathrm{k})}:\begin{Bmatrix} 0 \\ \boldsymbol{H}_{\mathbf{1}\mathrm{d}\varepsilon_{\mathrm{eq}}}^{(\mathrm{k})} \\ 0 \end{Bmatrix} - \frac{\partial f{:}\boldsymbol{A}^{(\mathrm{k})}:\begin{Bmatrix} \mathrm{d}\boldsymbol{\varepsilon}^{(\mathrm{k})} \\ 0 \\ 0 \end{Bmatrix} - \left(\mathrm{d}\varepsilon_{\mathrm{eq}}^{(\mathrm{k})}\right)^2 \partial f{:}\boldsymbol{A}^{(\mathrm{k})}:\begin{Bmatrix} 0 \\ \boldsymbol{H}_{\mathbf{1}\mathrm{d}\varepsilon_{\mathrm{eq}}}^{(\mathrm{k})} \\ 0 \end{Bmatrix}}{\partial f{:}\boldsymbol{A}^{(\mathrm{k})}:\left\{\tilde{\boldsymbol{n}}^{(\mathrm{k})}\right\}} \boldsymbol{A}^{(\mathrm{k})}:\left\{\tilde{\boldsymbol{n}}^{(\mathrm{k})}\right\} \tag{A.24}$$

A.2.2 Updating the Memory Surface

Within the modified Yoshida–Uemori hardening model, the memory surface as shown in Fig. A.2 is included to consider the non-isotropic hardening effect during reversal loading after the transient Bauschinger effect. This effect is reported to be quite obvious for some materials such as mild steel subjected to cyclic loading at large plastic reversals (Yoshida and Uemori 2002).

The evolution of the memory surface is updated at the end of the stress integration. The first-order Taylor series expansion of r^2 gives:

$$r^2 = r_n^2 + 2r \cdot \Delta r_n = r_n^2 + \frac{4h}{3}(\boldsymbol{\varepsilon}_p - \boldsymbol{q}):\Delta\boldsymbol{\varepsilon}_{p,n} \tag{A.25}$$

$$\boldsymbol{\varepsilon}_p - \boldsymbol{q} = \frac{\boldsymbol{\varepsilon}_p - \boldsymbol{q}_n}{1+\mu} \tag{A.26}$$

The memory surface

$$g_\varepsilon = \frac{2}{3}(\boldsymbol{\varepsilon}_p - \boldsymbol{q}):(\boldsymbol{\varepsilon}_p - \boldsymbol{q}) - r^2 = 0 \tag{A.27}$$

Substitute Eqs. (A.25) and (A.26) into Eq. (A.27), then

$$\frac{2}{3}\frac{(\boldsymbol{\varepsilon}_p - \boldsymbol{q}_n):(\boldsymbol{\varepsilon}_p - \boldsymbol{q}_n)}{(1+\mu)^2} - \left[r_n^2 + \frac{4h}{3}\frac{(\boldsymbol{\varepsilon}_p - \boldsymbol{q}_n):\Delta\boldsymbol{\varepsilon}_{p,n}}{1+\mu}\right] = 0 \tag{A.28}$$

$$\frac{2}{3}(\boldsymbol{\varepsilon}_p - \boldsymbol{q}_n):(\boldsymbol{\varepsilon}_p - \boldsymbol{q}_n) - r_n^2(1+\mu)^2 - \frac{4h}{3}(\boldsymbol{\varepsilon}_p - \boldsymbol{q}_n):\Delta\boldsymbol{\varepsilon}_{p,n}(1+\mu) = 0 \tag{A.29}$$

Then, μ can be obtained as

$$\mu = \frac{\frac{4h}{3}(\boldsymbol{\varepsilon}_p - \boldsymbol{q}_n):\Delta\boldsymbol{\varepsilon}_{p,n} + \sqrt{\left[\frac{4h}{3}(\boldsymbol{\varepsilon}_p - \boldsymbol{q}_n):\Delta\boldsymbol{\varepsilon}_{p,n}\right] + \frac{8}{3}r_n^2[(\boldsymbol{\varepsilon}_p - \boldsymbol{q}_n):(\boldsymbol{\varepsilon}_p - \boldsymbol{q}_n)]}}{2r_n^2} - 1 \tag{A.30}$$

Update the state variables related to the memory surface

$$\boldsymbol{q} = \boldsymbol{\varepsilon}_p - \frac{\boldsymbol{\varepsilon}_p - \boldsymbol{q}_n}{1+\mu} \tag{A.31}$$

$$r^2 = r_n^2 + \frac{4h}{3}(\boldsymbol{\varepsilon}_p - \boldsymbol{q}):\Delta\boldsymbol{\varepsilon}_{p,n} \tag{A.32}$$

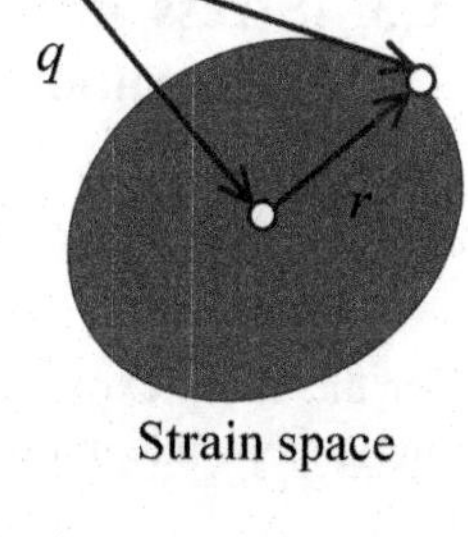

Fig. A.2 Illustration of memory surface proposed by Ohno

Update the bounding surface

$$\Delta R_n = mR_{\text{sat}}e^{-m\varepsilon_{\text{eq}}}\Delta\varepsilon_{\text{eq},n} + m_l\Delta\varepsilon_{\text{eq},n} \tag{A.33}$$

$$R = R_n + \Delta R_n \tag{A.34}$$

A.2.3 *Consistent Tangent Matrix of a Single-Step Integration Method*

For most of the implicit FE codes carry out iteration process at a global level using the backward Euler method, the consistent tangent matrix also called material Jacobian matrix in ABAQUS is required to be provided to carry out the iterative solution of the global equilibrium equations during the implementation of material constitutive models into an implicit FE code. It means that the implementation of the models concerns two aspects, i.e., integration of the state variables of the model and updating of the consistent tangent matrix at the end of the subroutine.

The consistent tangent matrix is very important especially for the material models with complicated formations. An inaccurate consistent tangent matrix will not only result in a slow convergence of the solution, but also sometimes can give a non-convergence result.

The consistent tangent matrix, $\boldsymbol{D}^{\text{alg}}$, is defined as

$$\boldsymbol{D}^{\text{alg}} = \frac{\partial \boldsymbol{\sigma}}{\partial \boldsymbol{\varepsilon}} \tag{A.35}$$

Based on Eq. (A.24), it can be found that the consistent tangent matrix which can be obtained at the end of the stress integration is the first 6 × 6 components of the following matrix

$$\boldsymbol{D}^{\text{alg}} = \boldsymbol{A} - \frac{(\partial f{:}\boldsymbol{A}) \otimes (\boldsymbol{A}{:}\{\tilde{\boldsymbol{n}}\})}{\partial f{:}\boldsymbol{A}{:}\{\tilde{\boldsymbol{n}}\}} \tag{A.36}$$

where ⊗ denotes the tensor product of two tensors.

A.3 Stress Integration of the Adaptive Substepping Integration Method

A.3.1 *Introduction of the Method*

For the complexity of the modified Yoshida–Uemori model, it was found that the single-step integration method as illustrated in Fig. A.1 cannot guarantee a convergence solution for the cases with a large plastic strain increment. In the present study, the substepping method as illustrated in Fig. A.3 was employed for the stress integration of the model to achieve a convergence solution. With this method, the stress integration process of the integration points failing to converge is divided into several substeps, and then, the non-convergence of the local integration problem can be solved successfully. Moreover, the number of substeps is controlled by using an adaptive technique to improve the computational efficiency of the local stress integration points with large strain increments.

The constitutive formulae are integrated from t_n to t_{n+1} in N substeps, from $t_{n+1/N}$ to $t_{n+N/N}$. During the stress integration of the $(n + 1)$th step, the stress, backstresses, and other state variables at the nth step are known, and strain increment at the $(n + 1)$th step is also known. The strain increment can be divided averagely into N substeps as

$$\mathrm{d}\boldsymbol{\varepsilon}_{n+1} = \mathrm{d}\boldsymbol{\varepsilon}_1 + \mathrm{d}\boldsymbol{\varepsilon}_2 + \cdots + \mathrm{d}\boldsymbol{\varepsilon}_N = \sum_{1}^{N} \frac{\mathrm{d}\boldsymbol{\varepsilon}_{n+1}}{N} \tag{A.37}$$

For the substepping integration method, the trial stress is first obtained using Eq. (A.1). The total number of the substeps, N, may be estimated as the closest integer of

$$N = \frac{\left|\boldsymbol{\sigma}_{n+1}^{\mathbf{trial}}\right|}{K\sigma_{y0}} \tag{A.38}$$

where $\left|\boldsymbol{\sigma}_{n+1}^{\mathbf{trial}}\right|$ and σ_{y0} are the elastic trial stress and initial yielding stress of the material, respectively, and K is a factor depending on the complexity of the

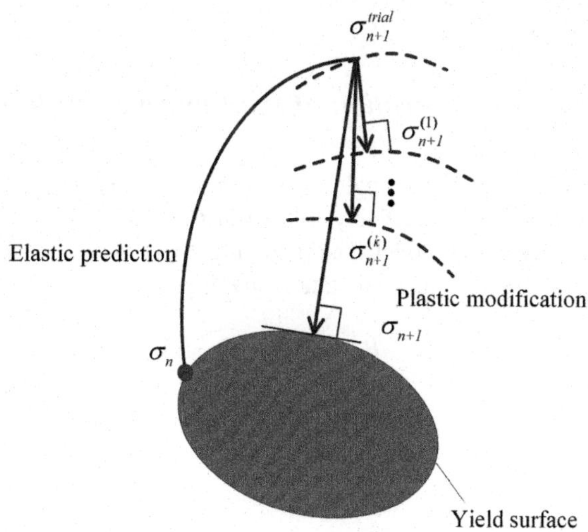

Fig. A.3 Illustration of adaptive substepping integration method

problem. This treatment can solve the non-convergence problem in some integration points with high strain concentration in extremely large plastic strain ranges. It is also reasonable to determine the number of substeps based on the value of the elastic trial stress which indicates the distance from the yield surface.

Once the number of substeps is decided, the solution can be accomplished by integration of each substep using the same procedure as that of the single-step integration method except that the strain increment is not $\mathrm{d}\varepsilon_{n+1}$ but $\mathrm{d}\varepsilon_{n+1}/N$, and all the state variables at the $(n + 1)$th step will then be obtained.

A.3.2 Consistent Tangent Matrix for the Adaptive Substepping Integration Method

The consistent tangent matrix at the end of the whole step has to be provided, which can be approximately obtained by the sum of all the consistent tangent matrix of the substeps. According to Eq. (A.35), the consistent tangent matrix of the whole step can be given as

$$\boldsymbol{D}^{\mathrm{alg}} = \frac{\partial \boldsymbol{\sigma}}{\partial \boldsymbol{\varepsilon}} = \frac{\sum_{i=1}^{N} \mathrm{d}\boldsymbol{\sigma}_i}{\sum_{i=1}^{N} \mathrm{d}\boldsymbol{\varepsilon}_i} \approx \sum_{i=1}^{N} \frac{\mathrm{d}\boldsymbol{\sigma}_i}{\mathrm{d}\boldsymbol{\varepsilon}_i} = \sum_{i=1}^{N} \boldsymbol{D}_i^{\mathrm{alg}} \tag{A.39}$$

where $\boldsymbol{D}_i^{\mathrm{alg}}$ is the consistent tangent matrix calculated at the end of the ith substep according to Eq. (A.35). The above formula is verified to be accurate enough for even large strain increment. It is also should be noted that sometimes $\boldsymbol{D}_i^{\mathrm{alg}}$ can be equal to the elastic modulus, $\boldsymbol{D}$, when the material does not yield at the first several substeps. Up until now, the implementation is completed with the stress integration process as shown in Table A.1.

A.3.3 Verification of the Implemented Model

Analyses using a single-element model under tension and shear with the boundary conditions, respectively, shown in Fig. A.4a and b, are carried out to validate the implemented model. The bounding stress, σ_{bound}, of the modified Yoshida–Uemori model under uniaxial state is given as

$$\sigma_{\mathrm{bound}} = B + R + \beta = B + (R_{\mathrm{sat}} + b)(1 - \mathrm{e}^{-m\varepsilon_{\mathrm{eq}}}) + m_l \varepsilon_{\mathrm{eq}} \tag{A.40}$$

Table A.1 Implementation scheme of modified Yoshida–Uemori model

1. Input data from the main FE code and set the initial values of the state variables as the converged values at the end of the previous step and compute trial stress For $k = 0$: $\varepsilon_p^{(0)} = \boldsymbol{\varepsilon}_{p,n}$, $\boldsymbol{\alpha}^{(0)} = \boldsymbol{\alpha}_n$, $\boldsymbol{\theta}^{(0)} = \boldsymbol{\theta}_n$, $\boldsymbol{\beta}^{(0)} = \boldsymbol{\beta}_n$, $\boldsymbol{q}^{(0)} = \boldsymbol{q}_n$, $\boldsymbol{\sigma}_{n+1}^{\text{trial}} = \boldsymbol{\sigma}_n + \boldsymbol{D}{:}\mathrm{d}\boldsymbol{\varepsilon}_{n+1}$

2. Check yield condition
If $\sigma_{n+1}^{trial} - \sigma_{y0} < 0$, then
Set $(*)_{n+1} = (*)_{n+1}^{\text{trial}}$ and exit the local integration and return the main FE code
End if
Otherwise, calculate the total number of substeps and flow direction n_{n+1}
$\boldsymbol{n}_{n+1} = \boldsymbol{\sigma}_{n+1}^{trial} / \sigma_{n+1}^{trial} \quad N = \sigma_{n+1}^{trial} / \sigma_{y0}$
and go to step 3

3. Compute equivalent plastic strain increment using the backward-Euler iteration
4. Update backstresses, plastic strain and stress, and equivalent plastic increment $\begin{Bmatrix} \mathrm{d}\boldsymbol{\sigma}^{(k)} \\ \mathrm{d}\boldsymbol{\theta}^{(k)} \\ \mathrm{d}\boldsymbol{\beta}^{(k)} \end{Bmatrix} = -\left[\boldsymbol{A}^{(k)}\right]{:}\left\{\tilde{\boldsymbol{F}}^{(k)}\right\} - \delta \mathrm{d}\varepsilon_{eq}^{(k)} \left[\boldsymbol{A}^{(k)}\right]{:}\left\{\tilde{\boldsymbol{n}}^{(k)}\right\} - \left(\mathrm{d}\varepsilon_{eq}^{(k)}\right)^2 \left[\boldsymbol{A}^{(k)}\right]{:}\left\{\tilde{\boldsymbol{H}}_{\mathbf{1}\mathrm{d}\varepsilon_{eq}}^{(k)}\right\}$
5. Compute consistent tangent matix of substep $\boldsymbol{D}_i^{\text{alg}} = \boldsymbol{A} - \frac{(\partial f{:}\boldsymbol{A}) \otimes (\boldsymbol{A}){:}\{\tilde{\boldsymbol{n}}\})}{\partial f{:}\boldsymbol{A}{:}\{\tilde{\boldsymbol{n}}\}}$ Repeat Step 3–5 from the first substep to the Nth substep
6. Update the consistent tangent matrix of the whole step by summation of all the consistent tangent matix of substeps $\boldsymbol{D}^{\text{alg}} = \sum_{i=1}^{N} \boldsymbol{D}_i^{\text{alg}}$

Theoretically, the stress (equivalent Mises stress) obtained from the numerical results should start from the yield stress and gradually converges to the bounding stress given in Eq. (A.40) as the plastic strain increases. The materials DP600 (Ghaei et al. 2010) and SS400 (Jia and Kuwamura 2014) with the model parameters given in Table A.2 are employed to validate the implemented subroutine. The comparisons between the numerical results and the corresponding theoretical ones are, respectively, given in Fig. A.5a and b for DP600 and SS400. It can be found that the stresses of the numerical results converge to the theoretical bounding stresses as strain increases.

A.3.4 *Effect of Substep Length on the Robustness and Computational Efficiency*

Analyses using a single-element model similar to those shown in Fig. A.4a and b are carried out to investigate the effect of substep length, $K\sigma_{y0}$, on the robustness and computational efficiency of the implemented model, where cyclic displacement loadings are applied to the single-element model. Small values are defined for the output numbers of increments to produce large plastic strain increments during the

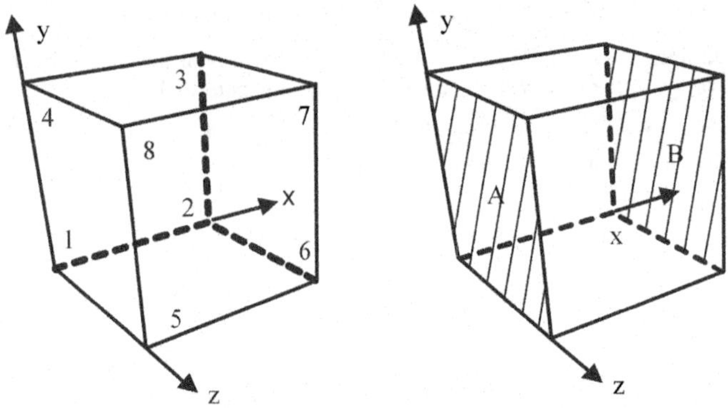

Initial boundary conditions:
Node1 : Ux=Uy=Uz=0;
Node2 : Uy=Uz=0;
Node3 : Uz=0;
Node4 : Ux=Uz=0;
Node5 : Ux=Uy=0;
Node6 : Uy=0

Applied displacement
Nodes5 , 6 , 7 , 8 : Uz

(a) Single solid element under tension

Initial boundary conditions:
Surface A: Ux=Uy=Uz=0;
Surface B: Ux=0;

Applied displacement
Surface B : Uy

(b) Single solid element under shear

Fig. A.4 Calibration of the plasticity model using a single-element model

integration. The CPU times of the analyses using the implemented model with different substep lengths and the one using the single-step integration method are compared in Fig. A.6. The CPU time T is non-dimensionalized by the one using the substepping integration method with a substep length of σ_{y0}, T_1, and the investigated substep length ranges from $\sigma_{y0}/4$ to 10 σ_{y0}. For the single-element model under cyclic axial loading, it is found that the optimal substep length is about 5 σ_{y0}, while equals to σ_{y0} for the case under shear loading. The corresponding non-dimensionalized CPU time of the one using the single-step integration method is also presented in the figure for the case under shear loading, indicating that the substepping integration method with a substep length of σ_{y0} can save 16% CPU time compared with the single-step integration method. The CPU time of the case under cyclic axial loading using the single-step integration method is not shown in

Table A.2 Parameters of the modified Yoshida–Uemori model

Materials	σ_{y0}	C	B	R_{sat}	b	m	h	m_l	$\varepsilon_{plateau}$
DP600	420.0	200.0	555.0	190.0	110.0	12.0	0.9	0.0	0.0000
SS400	255.9	332.8	321.7	137.7	82.9	18.1	0.5	236.2	0.0148
STKR400	233.2	369.6	261.0	139.4	295.0	2.8	0.1	206.0	0.0222

Notes Unit of σ_{y0}, C, B, R_{sat}, b, and m_l: MPa
Unit of m, h, and $\varepsilon_{plateau}$: dimensionless

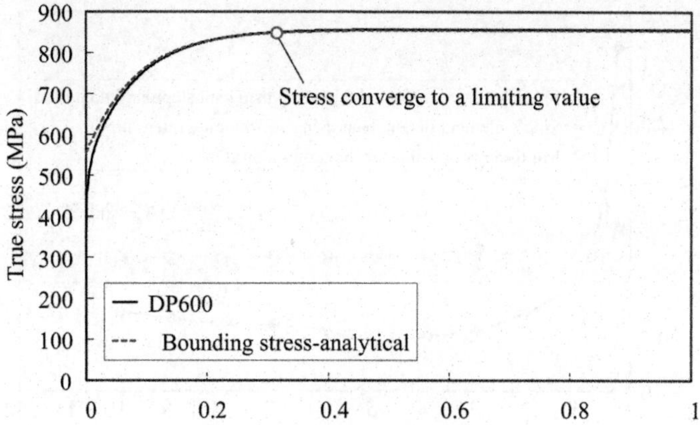

(a) Material DP600

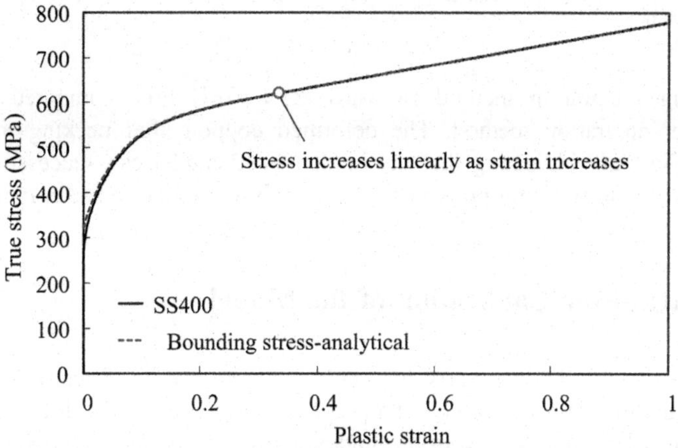

(b) Material SS400

Fig. A.5 Calibration results using single-element model

the figure, for the numerical analysis does not converge due to the large strain increment. However, the cases using the substepping integration method all converge, and the optimal substep length is about σ_{y0}. This implies that the substepping integration method can solve those large strain increment problems which cannot converge using the single-step integration method. A multi-element model (a one-fourth model for a coupon test) as shown in Fig. A.7 is also investigated to study the effect of substep length in more complicated loading conditions. The non-dimensionalized CPU times using the substepping integration method with different substep lengths and the single-step integration method are all given in Fig. A.6, indicating that the optimal substep length is around σ_{y0}, and the

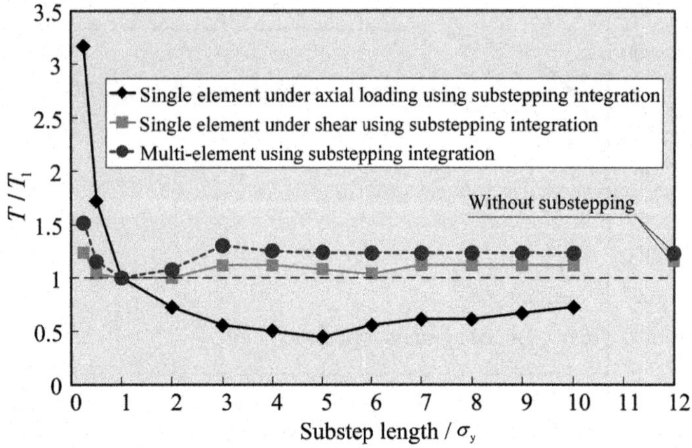

Fig. A.6 Robustness of substepping integration method and effect of substep length on computational efficiency

substepping integration method can save 23% CPU time compared with the single-step integration method. The deformed coupon after necking initiates is given in Fig. A.8, indicating that the implemented model can successfully solve severe strain concentration problems at extremely large plastic strain ranges.

A.4 Parameter Calibration of the Model

In practice, it is difficult to carry out cyclic coupon tests till fracture due to premature buckling at compressive half cycles. However, it is found (Jia and Kuwamura 2014) that the isotropic and kinematic hardening components each approximately takes half of the hardening stresses obtained from a monotonic tension coupon test. Based on this rule, one can obtain cyclic stress–strain data from a monotonic coupon test result. During the calculation of the true stress–true strain data under monotonic tension, one has to modify the true stresses after necking initiation, and the detailed method has been presented in Chap. 2. When the cyclic stress–strain data is obtained, an optimization analysis using the single-element model as shown in Fig. A.4a will be carried out, where an optimization analysis software such as MATLAB is required to fit the analysis results from ABAQUS with the cyclic true stress–true strain data constructed from the monotonic ones. The parameters with the best fitting with the constructed cyclic stress–strain data are the model parameters for the modified Yoshida–Uemori model.

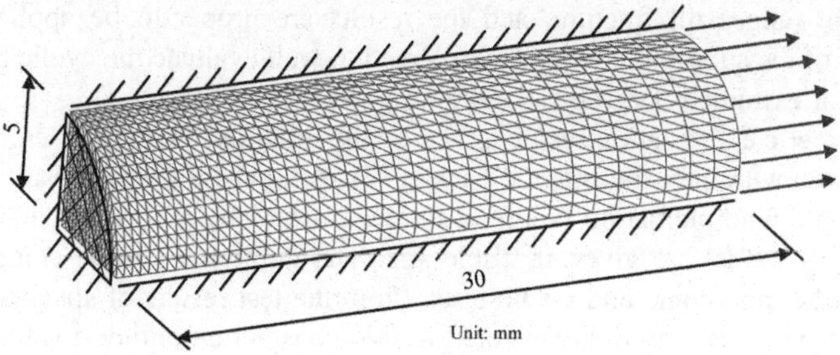

Fig. A.7 Calibration of the plasticity model using multi-element model

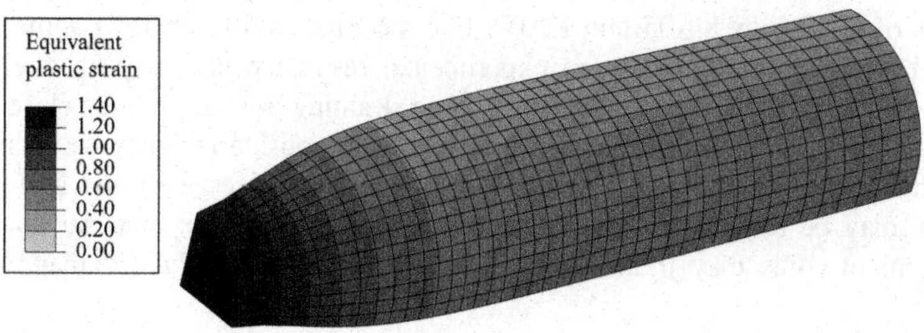

Fig. A.8 Calibration results using the multi-element model

A.5 Application of the Model to Extremely Large Plastic Straining Problems

The modified Yoshida–Uemori model is incorporated into ABAQUS/Standard, and results of a series of cyclic tests on mild steels until fracture at both material and member levels, respectively, given in Chaps. 3 and 7 are employed to compare with the numerical results and further validate the applicability of the model. Cyclic tests on hourglass specimens under compression and tension were carried out under seven different loading histories, and the material is SS400. The specimens KA01 and KA02 were respectively subjected to monotonic tension and compression. specimen KA03 was subjected to a single full-cycle loading. Specimen KA04 was first under five constant-amplitude loading cycles and then pulled to fracture. Specimen KA05 was first subjected to two constant-amplitude cycles before necking of the specimen initiated. Specimen KA06 was cyclically loaded at the strain ranges beyond the necking initiation strain, where the necking initiated at a strain about 20%. Specimen KA07 was first cyclically loaded within strain ranges smaller than the necking initiation and then within the ranges larger than the necking initiation strain. All the specimens were finally pulled to fracture except for specimen KA02, which failed prematurely due to buckling. The tests covered broad large strain ranges till fracture, and the results are proper to be applied to the validation of the subroutine, which is expected to well evaluate the cyclic behaviors of metals at extremely large plastic strain ranges.

The numerical simulation results given by the modified Yoshida–Uemori model are compared with the experimental ones in Fig. A.9, where the test result of KA02 is not given due to premature failure of the specimen. The calibrated parameters of the plasticity model are given in Table A.2. The numerical results compare well with both the monotonic and cyclic tests. From the test results of specimen KA04 where the specimen was first subjected to five constant-amplitude loading cycles, the forces in the experiment tend to stabilize after the second loading cycle, and the numerical results can well describe the non-isotropic hardening phenomenon after the second loading cycles. The model also can well describe the cyclic plasticity of the mild steel at extremely large plastic strain ranges, which can be found in the test results of specimens KA05 and KA07. For specimen KA06, there is some deviations between the numerical and experimental results, which may be due to the deterioration of the material under the cyclic straining at strain ranges close to the fracture strain. At this stage, a constitutive model considering damage of metals, e.g., the Gurson model or GTN model (Gurson 1975; Tvergaard and Needleman 1984), may be required to give an accurate prediction of the material plasticity, since micro-voids may initiate within metals which will induce damage of the material.

Two square hollow section stub columns with heat treatment, respectively, under cyclic incremental and constant-amplitude loading are employed for the calibration. Local buckling first occurred in both of the columns, and post-buckling ductile fracture initiated in subsequent loading cycles. The model parameters are calibrated from the corresponding coupons cut from the same tube. A one-eighth model with the symmetric boundary conditions shown in Fig. A.10 is established, and cyclic displacement loadings are applied to the top end. The comparisons between the numerical and experimental results are given in Fig. A.11, and the crack initiation points are also marked in the curves. It is indicated that the plasticity model can well evaluate the load–displacement curves of the stub columns under both incremental and constant-amplitude loadings before cracking occurs and overestimates the loads after cracking. To trace the load–displacement curves after cracking, the crack propagation has to be simulated to consider the loss of strength due to the damage. The local buckling of the specimens is also well predicted by the simulation as illustrated in Fig. A.12.

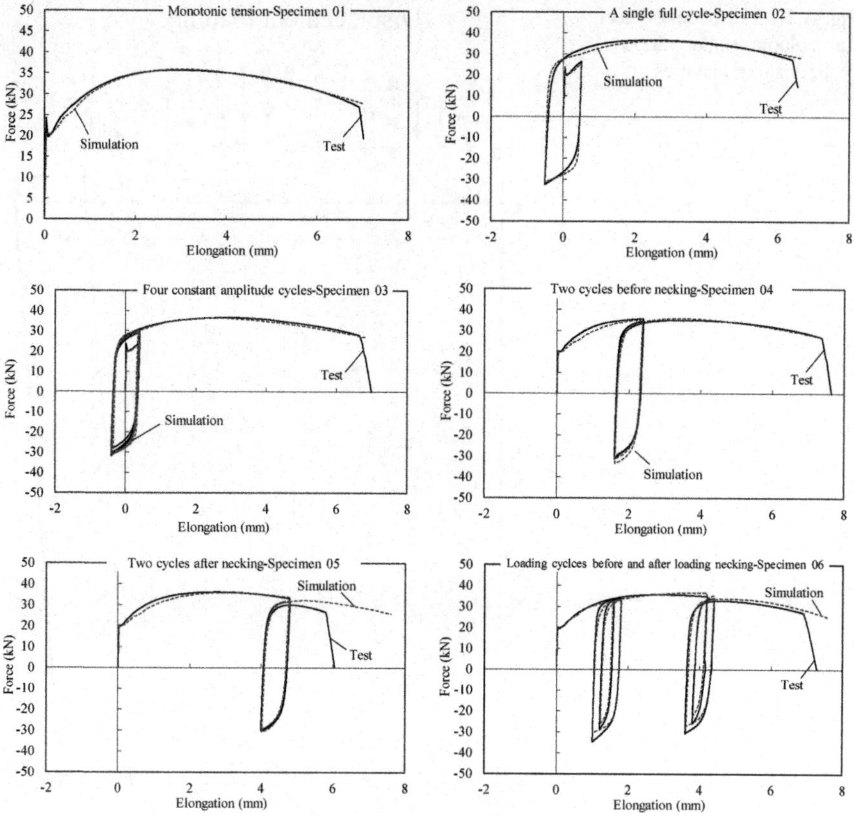

Fig. A.9 Application to structural steel under various cyclic loading histories

A.6 Summaries

An adaptive substepping integration algorithm is employed to implicitly integrate the modified Yoshida–Uemori model. The detailed integration process is given, and the integration equations are linearized and solved using the backward Euler iteration method. A method termed as adaptive substepping integration method is also proposed to determine the proper number of the substeps in a large plastic strain increment, and the number of the substeps is related to the value of the elastic trial stress, which represents the distance away from the yield surface. The optimal substep length of the implemented modified Yoshida–Uemori model is found to be around σ_{y0} for highly strain concentration problems. The integration algorithm is

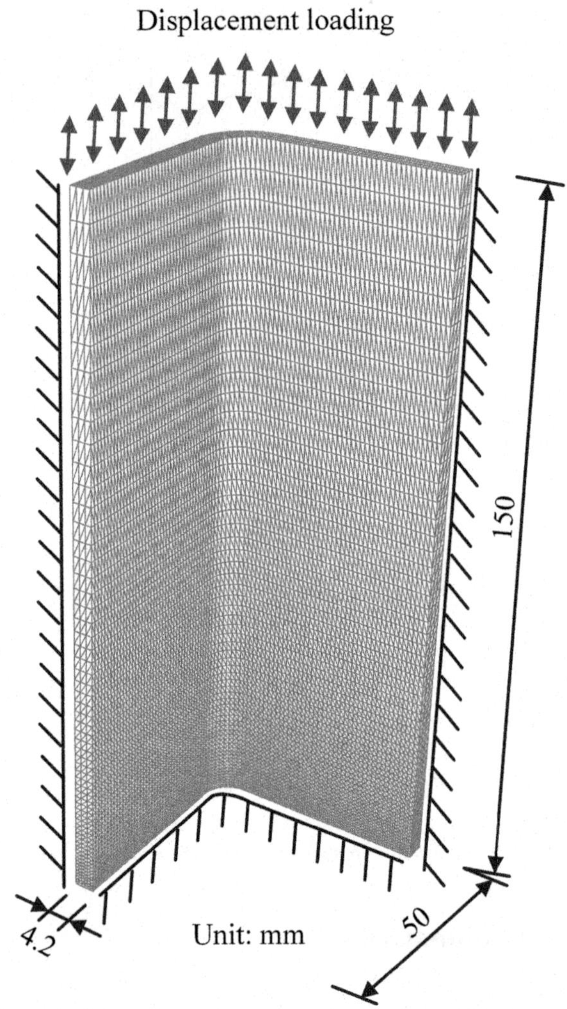

Fig. A.10 FE model of steel stub columns under various cyclic loading histories

found to be robust even at extremely large plastic strain ranges close to fracture strains of metals and also effective in both robustness and computational speed compared with the single-step integration method. Finally, the subroutine was further validated with theoretical results and cyclic test results of mild steels at both material and member levels. The numerical analyses can converge up to fracture of the material and stub columns, and the numerical results also compare with the experimental ones with good accuracy. The modified model is also further validated

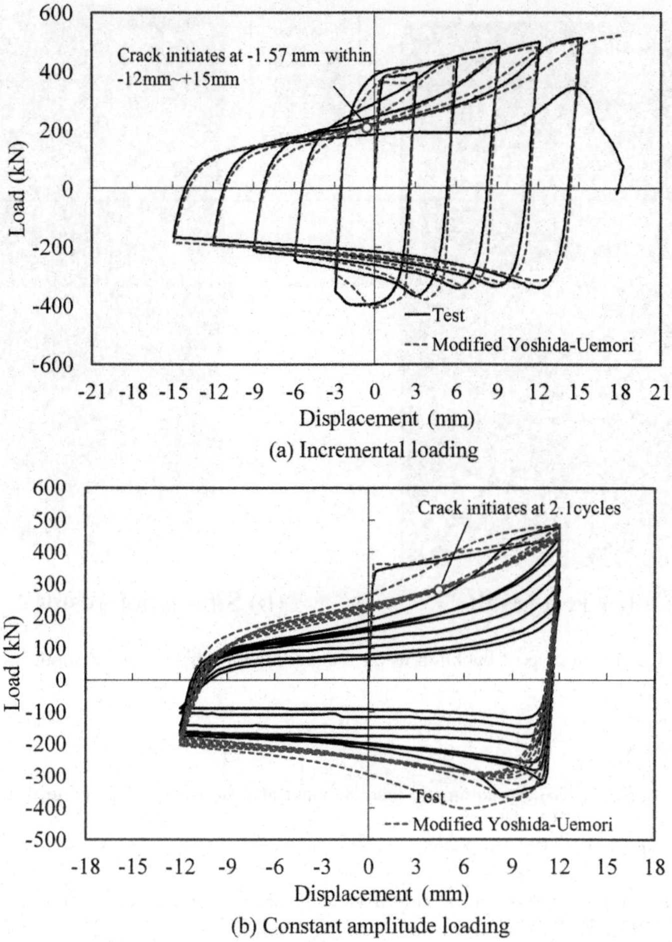

Fig. A.11 Application of the modified Yoshida–Uemori model to steel columns under cyclic large strain loading

by cyclic tests of steel members with strain concentration at the buckled regions. The algorithm is applicable to a range of constitutive plasticity models with complicated formations, which are difficult to converge at extremely large plastic strain.

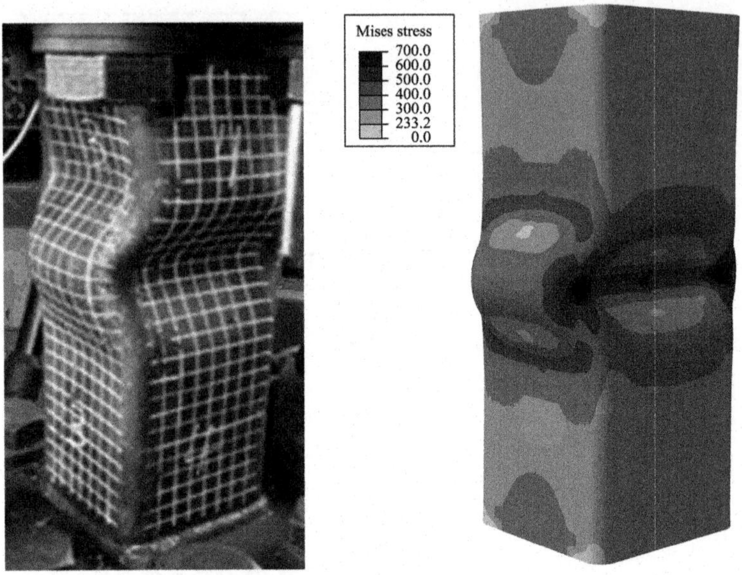

(a) Test result (b) Simulation result

Fig. A.12 Simulation of post-buckling using the modified Yoshida–Uemori model

References

Abbo AJ, Sloan S (1996) An automatic load stepping algorithm with error control. Int J Numer Methods Eng 39:1737–1759

Bao Y, Wierzbicki T (2004) On fracture locus in the equivalent strain and stress triaxiality space. Int J Mech Sci 46:81–98

Bao Y, Wierzbicki T (2005) On the cut-off value of negative triaxiality for fracture. Eng Fract Mech 72:1049–1069

Bićanić N, Pearce C (1996) Computational aspects of a softening plasticity model for plain concrete. Mech Cohesive—Frictional Mater 1:75–94

Crisfield M et al (2012) Nonlinear finite element analysis of solids and structures. Wiley

Dunne F, Petrinic N (2005) Introduction to computational plasticity. Oxford University Press, Oxford

Ghaei A, Green DE (2010) Numerical implementation of Yoshida-Uemori two-surface plasticity model using a fully implicit integration scheme. Comput Mater Sci 48:195–205

Ghaei A et al (2010) Semi-implicit numerical integration of Yoshida-Uemori two-surface plasticity model. Int J Mech Sci 52:531–540Gurson AL (1975) Continuum theory of ductile rupture by void nucleation and growth. Part I. Yield criteria and flow rules for porous ductile media. Brown University, Providence, RI (USA). Division of Engineering

Jia L-J, Kuwamura H (2014) Ductile fracture simulation of structural steels under monotonic tension. J Struct Eng (ASCE) 140:04013115

Jia L-J, Kuwamura H (2014) Prediction of cyclic behaviors of mild steel at large plastic strain using coupon test results. J Struct Eng (ASCE) 140:04013056

Jia L-J (2013) Ductile fracture of structural steels under cyclic large strain loading. Doctor, The University of Tokyo, Tokyo

Ohno N (1982) A constitutive model of cyclic plasticity with a nonhardening strain region. J Appl Mech 49:721–727
Pérez-Foguet A et al (2001) Consistent tangent matrices for substepping schemes. Computer methods in applied mechanics and engineering 190:4627–4647
Pérez-Foguet A, Armero F (2002) On the formulation of closest—point projection algorithms in elastoplasticity—part II: Globally convergent schemes. Int J Numer Methods Eng 53:331–374
Shi MF et al (2008) Determination of nonlinear isotropic/kinematic hardening constitutive parameters for AHSS using tension and compression tests. In: Proceedings of NUMISHEET Conference, Switzerland, 264–270
Simo J, Hughes T (1998) Computational inelasticity. New York
Sloan S (1987) Substepping schemes for the numerical integration of elastoplastic stress–strain relations. Int J Numer Methods Eng 24:893–911
Tvergaard V, Needleman A (1984) Analysis of the cup-cone fracture in a round tensile bar. Acta Metall 32:157–169
Wierzbicki T et al (2005) Calibration and evaluation of seven fracture models. Int J Mech Sci 47:719–743
Yoshida F, Uemori T (2002) A model of large-strain cyclic plasticity describing the Bauschinger effect and workhardening stagnation. Int J Plast 18:661–686